THE SURVIVAL GAME

THE
SURVIVAL
GAME

How Game Theory
Explains the Biology of
Cooperation and Competition

David P. Barash

TIMES BOOKS
Henry Holt and Company • New York

Times Books
Henry Holt and Company, LLC
Publishers since 1866
115 West 18th Street
New York, New York 10011

Library of Congress Cataloging-in-Publication Data
Barash, David P.
 The survival game : how game theory explains the biology of
cooperation and competition / David P. Barash.
     p.   cm.
 Includes bibliographical references and index.
 ISBN: 0-8050-7175-X
 1. Social interaction. 2. Cooperativeness. 3. Competition (Psychology)
4. Choice (Psychology) 5. Game theory. I. Title: Game theory explains the
biology of cooperation and competition. II. Title.
HM1111.B37 2003
302—dc21   2003054351

First Edition 2003

Printed in the United States of America

1   3   5   7   9   10   8   6   4   2

"It is not a question only of simple games but . . . the foundation is being laid for interesting and deep speculations."

—Christian Huygens (1629–1695),
Dutch mathematician,
physicist, and astronomer

"Any event . . . may be regarded as a game of strategy, if one looks at the effect it has on the participants."

—John von Neumann (1903–1957),
Hungarian-born American mathematician
and one of the founders of game theory

CONTENTS

THE SURVIVAL GAME

1

The Games We All Play:
What They Are, Why They Matter

In Molière's play *Le Bourgeois Gentilhomme,* Monsieur Jourdain is astonished to learn that all of his life, without knowing it, he has been speaking prose. We are a bit like M. Jourdain: without knowing it, we all play games. It is not necessary to be athletic, or competitive, or especially frolicsome. Game playing is a big part of life. And since we are full-time players, it behooves us to understand what's going on.

Here goes.

What's the Big Idea?

Most of us assume that life is straightforward, essentially under our own control. If we want something, and reach for it, we may succeed or fail. Either way, the outcome is widely thought to result from our actions alone. But in fact, what we get is often determined by factors out of our control: Maybe the object we are reaching for is too heavy, or too far away, or guarded by angry dragons.

For our purposes, there is a whole class of situations that are more complex and more interesting than these, circumstances in which the payoff—the gain we are seeking—is limited by the fact that others are also reaching for the same goal. In cases of this sort, as the Rolling Stones proclaimed in a notable song several decades ago, you can't always get what you want. Why not? Because if someone else wants the same thing, and if he or she is pretty much as smart, fast, strong, and motivated as you, then something's got to give. (Whether, in the end, you can get "what you need," as the Stones also announced, is another question, and one that is even more complicated.)

Actually, games arise even if the players aren't human beings. Animals also play games, whether they know it or not, just as people do—whether *they* know it or not. Thus, two bull elk may desire the same cow, with the success of each ultimately depending not just on what male number one does, but also on male number two. And, of course, the female is also likely to have something to say about the outcome. Whether people or animals (or even viruses), the important thing is that there is some sort of outcome, which is determined by the combined actions of two or more different players, whether their interests are shared, opposed, or—most commonly—a little bit of each.

Let's get a bit more technical, but not much. There are many circumstances in which the interests of individuals are interdependent and yet in conflict, so that the payoff to individual A, who is pursuing a particular goal, depends on the actions of individual B, who may be pursuing the same goal. In these cases, the return to each player—which can be a person, animal, organization, country, or even a bacterium—is determined by the actions of both, taken together. Furthermore, each is in a sense at the mercy of the other, in two ways. First, the outcome to each party depends on the other's actions, and second, it is often the case that neither can change the other's behavior. It is one of life's crucial constraints.

There are many variants on this theme. Consider, for example, one of the simplest: call it the Interrupted Telephone Call Game, a frustration that everyone has experienced. You are talking to a friend, long distance, and suddenly the conversation is cut off; you both get a dial tone. You want to resume talking, and so does your friend, but if you both dial up the other, neither one will get through! If you both wait, again you both lose. The only way to "win" is for one (either one) to redial and the other to wait. So this is a case in which the interests of both parties converge, and yet the payoff to each *still* depends on the independent behavior of the other.

Another way to put it: What is the best thing to do when confronted with the Interrupted Telephone Call Game? There is no simple answer here, since it depends on what the other person does. If she is going to call you, then you should wait. If she is going to wait, then you should call. Things get interesting when—as is usually the case—the two of you haven't agreed in advance who will do what if your call is interrupted.

Although this is admittedly a trivial case, it points out something important about interactions of this sort: Two or more parties may each have a limited number of possible "moves"—in this case, you and your friend can either "dial" or "wait"—with the payoff to each of you depending not just on what either of you do, but on what the other does at the same time. Moreover, neither can control the actions of the other. A poignant literary example comes from O. Henry's short story "The Gift of the Magi," in which the young husband sells his cherished pocket watch in order to buy hair combs for his adored wife, who—independently—has sold her precious and much-admired hair to buy a watch chain for him!

In some of the most interesting situations we'll be examining, the "players" are more competitive, if not overtly antagonistic. In such cases, each participant is typically trying to *maximize* his payoff while often simultaneously attempting to *minimize* the other's return. (And each also knows that the other is trying to achieve the same thing: "She knows that I'm trying to seem smarter than her, so she'll probably study this other case carefully, to get a jump, so I'll review this counter-example, to get ahead of her. . . .") Interestingly, although experts in game theory typically assume this sort of conscious planning and counter-planning, it isn't strictly necessary. A gazelle "knows" that the cheetah is planning to catch it, and the cheetah "knows" that the gazelle "knows" this, and so forth. In this case, the "knowledge" is implanted by evolution rather than by conscious awareness, but nonetheless, the game goes on.

Competitive interactions of this sort, whether human or "merely" biological, are not only intriguing as intellectual exercises, but they can yield tremendous insight into important real-life dilemmas, whether inter-personal or involving whole societies. In other cases, two individuals—

or companies, or countries—find themselves locked in a deeply frustrating dilemma, in which both "players" strive for their own best interest, but, as a result, both are worse off. This is not simply theory but, rather, painful and dangerous practice.

Take, for example, nuclear weapons in Pakistan and India. Each country is tempted by the prospect of gaining a nuclear advantage over the other; at the same time, each would be better off using its limited budget to enhance the welfare of its own impoverished people. But each country is also fearful of being taken advantage of by the other if it lets down its guard and forgoes nuclear weapons. And so, two countries that can ill afford such a dangerous and expensive competition find themselves locked in a nuclear arms race that does neither one any good . . . and that, moreover, does harm to their own security and that of the rest of the world. Everyone would be better off if these two "players" would only "do the right thing" and stop their nuclear competition, but because each fears being suckered by the other, both see themselves as doomed to keep it up. As we'll see, arms races of this sort also occur between married couples, parents and children, and so on.

Once again, the biological world fits right in, although such "natural" arms races are less potentially lethal than their nuclear counterparts. Pity the poor peacock, for example, forced to grow a set of outlandish and metabolically expensive tail feathers, which threaten to get tangled in the undergrowth and serve no real purpose other than convincing the female that its possessor is better than his rivals. After all, if a particular peacock decided not to run this tail-feather race, such a presumably "rational" decision would place him at a disadvantage relative to the other males who decided to participate, and who, as a result, got the peahen.

Even trees are victims. Given that successful reproduction is the biological bottom line, why should redwoods grow so tall? After all, you don't have to be two hundred feet in height, and bother piling up hundreds of tons of wood, just to make some tiny seeds. But a redwood tree that opted out of the big-and-tall competitive fray would literally wither in the shade produced by other trees that were just a bit less restrained. And so, redwoods are doomed by their own unconscious selfishness to be "irrationally" large, for no particular reason other than the fact that other redwoods are doing the same thing.

Then there is the "politician's dilemma" of whether or not to "go negative." By doing so, someone running for public office doesn't merely have to invest in dead wood; he or she also loses respect and society loses the opportunity to debate genuine issues. But political rivals, not unlike aspiring redwood trees, often find themselves stuck in an awkward competitive game, in which they typically fear being suckered by their opponent (victimized this time by shady, negative campaign tactics), as well as tempted to reap the benefits of attacking successfully and unilaterally. Or like two contestants in a particularly grueling tug-of-war, each side may long to ease up, but fears that the other will take advantage, so both sides end up holding tight, straining mightily . . . and often getting nowhere. Not uncommonly, the two players come out somewhat behind, the only winners in the world of electoral politics being the consultants, the speechwriters, and the media.

Don't miss the forest for the trees (and not only redwoods). There are some shared threads linking situations of this sort, from gargantuan redwoods, horny peacocks and elk, to interrupted telephone calls, negative political ads, and dilemmas of disarmament. In all these cases, two sides, or players, each have goals or potential payoffs they wish to attain. They each have a limited palette of options, things they can do in pursuit of their goals; in the simplest case, just two (grow big—or fancy—or not, dial or wait, arm or disarm, go negative or stay positive). And although each is free to choose what to do, no one is free to obtain the desired payoff simply by reaching for it. In each case it depends on what the other guy does. Get used to it: you can't always get what you want. Especially if someone else's desires interfere with your own.

There is a complex branch of mathematics that handles such situations. Known as *game theory*, it has been around for about sixty years.* Although it has generated hundreds of scholarly articles (and

---

*This takes the publication of John von Neumann and Oskar Morgenstern's masterpiece, *Theory of Games and Economic Behavior* (1944), as the starting point; alternatively, one could argue for "more than seventy years," beginning with von Neumann's publication of his proof of the minimax theorem (1928).

several technical journals devoted entirely to its analysis) as well as many academic books, game theory has never truly reached the general public.* This is a shame, because it offers many rewards. As I hope to show, it provides a novel and intellectually compelling way of looking at everyday phenomena. And the fact that the same principles apply to the unthinking, biological world suggests that game theory itself may be in touch with some deeper truths: not a "theory of everything," as some physicists have been pursuing, but at least a theory of many interesting things. In addition, I believe that its essence can be conveyed without elaborate mathematics. In fact, I hereby promise to make this book an equation-free zone. Game theory and its implications are simply too important—and too much fun!—to be left to the mathematicians. Our games, our selves.

Games are also too dangerous to be ignored. Consider the Game of Chicken, a version of which was memorably portrayed in the James Dean movie *Rebel Without a Cause*. In classic Games of Chicken, two cars head toward each other, each daring the other to swerve. The one who chickens out is the loser; the one who is so brave, or so stupid, as to persevere in going straight is the winner. (In the movie version, Dean and his rival each drove toward a cliff, seeking to be the last to bail out; the basic principle remains the same.)

A Game of Chicken is similar to a nuclear confrontation in that each player can either insist on pushing straight ahead ("arm") or swerve ("disarm"). Moreover, each would get the highest payoff by doing the former if at the same time the other did the latter. On the other hand, the Game of Chicken differs from an arms race in one crucial respect. The worst outcome for either player in the latter case arises if its side acted cooperatively while the other acted competitively; by contrast, the worst outcome in the Game of Chicken occurs when *both* players act competitively (each hoping that the other will "swerve"). The Cuban Missile Crisis in 1962 was a terrifying example of international Chicken, in which the Soviet Union swerved ... thereby

---

*There are a few exceptions, notably Sylvia Nasar's book, *A Beautiful Mind* (Simon & Schuster, 1998) and the 2001 movie with the same title. Book and movie deal with the life of John Nash, one of the pioneers of game theory, and both are especially concerned with his decades-long struggle with schizophrenia, hardly touching on game theory itself.

avoiding fried chicken. There are many situations in daily life—issuing "take it or leave it" ultimatums, for example—when we engage in lower-risk Games of Chicken, usually without realizing it.

Biologists have begun to identify "game theoretic" strategies by which animals—and not just genuine chickens—play, quite seriously, at survival, at bluffing, and in reproductive roles. Of course, they don't need to know, consciously, that they are playing such games, any more than they need to understand the details of digestive physiology in order to eat.

Take this example. A male bluebird must "decide" between two options: He can remain with his sexually receptive mate, or wander off in search of additional female companions. He cannot do both, just as individuals playing the Game of Chicken cannot both go straight and swerve, or two countries caught in a nuclear arms race cannot both arm and disarm. Similarly, the male bluebird's payoff depends on what others are doing: If most other males are staying home, he may do well by looking for additional sexual opportunities, because there is at least the chance that some females will be left unguarded, and little risk in trying. But if too many other males are also looking to sow their wild bluebird oats, then a gallivanting male runs the risk that while he is out seeking copulations, another male—doing the same as himself—will succeed in copulating with *his* female! The apparent result, understood via game theory, is that individuals are likely to be either homebodies or sexual adventurers, in predictable proportions, depending on the risks, the payoffs, and what others in the population are up to.

Ironically, game theory may even apply *more directly* to the interactions of animals than of people, despite the fact that animals are by all accounts less rational than human beings. This is because animals are more driven by automatic processes, the results of natural selection having endowed animals with automatic responses that have largely proven, over many generations, to be fitness-enhancing and, thus, mathematically valid. By contrast, human beings are less automatic and, thus, *less* logically predictable. Here is yet another fundamental paradox, at the heart of all behavior, human and nonhuman alike.

This is but scratching the surface. Game theory is loaded with implications for a wide range of human activities; it offers intriguing mental exercise plus the insight that comes from seeing old problems in new ways. By examining interactions and the striving for payoffs in a "gamey" way, it is possible to shed new light on many other situations, including military strategy, stock market investing (buying a stock is a good strategy if and only if others will also buy it; you can't make money on Wall Street based on what you do alone), as well as moral decision making. And lots more.

We'll also use game theory as a lens to focus on interactions between individuals as well as organizations. Again, the underlying commonality is simply this: what one "player" gets isn't determined simply by what he, or she, or it does, but also by the other "player," who is no less smart, no more ethical, and every bit as motivated to succeed. Imagine this: You are accosted by your neighbor, because one of your trees fell down on her prize rosebush. She threatens to sue, yet both of you would rather settle out of court. She would like to get as much money as possible; you would like to give up as little as you can get away with. She must decide whether to demand a large amount of money, or a smaller amount. Independently, you must decide whether to agree or take it to your insurance company and, possibly, small claims court . . . which would be expensive and time-consuming for all concerned.

Your rose-fancying neighbor doesn't want to ask for too much—not wanting to drive both of you to serious litigation—but neither does she want to settle for too little. Similarly, you don't want to make it too clear at what point you'd settle, or else Rosey will ask right up to that amount. It's a cat-and-mouse game, which, in fact, is not a bad way to describe much of game theory, and also much of life.

While writing this chapter, I was invited by a large and well-funded organization to give an after-dinner lecture. How much should I request as my fee? If I asked too little, I'd be underpaid; if I demanded too much, I might drive them to ask someone else. So I wasn't free to simply ask for a lot of money, because I might be stuck with none at all. My "move" was determined by my awareness that the organization had "moves" of its own: it could agree, or refuse, leaving me high and dry. (I also knew, to my advantage, that their original choice had just withdrawn because of ill health, and the meeting was a mere two weeks away; I didn't quite have them "over a barrel," but this information gave me some helpful, added leverage.)

The roots of game theory go very deep, certainly deeper than the formal mathematics itself (not to mention your neighbor's roses). If, as seems likely, people have been playing "games" for millennia, all this gamesmanship has doubtless left its imprint on our very natures. And so, we assess prospects and possibilities, and are remarkably good intuitive statisticians. We spend time and energy seeking to "read" each other, playing "what if" games in our heads, often without even realizing it. We may even owe one of our most cherished traits—consciousness itself—to the game-playing propensities of our ancestors. The likelihood is that consciousness evolved because it leads to self-awareness; being conscious is, to a large extent, not just being aware, but being aware of being aware. And why is such self-awareness useful? Because it gives us the opportunity to assess what someone else may be thinking and feeling, and, thus, makes us better game players: "If she does that, then I ought to do this . . ." Or even further: "She's probably thinking that I'm going to do this, so she's likely to do that, so maybe I'll fool her by . . ."

Let's not get too carried away with this idea, however, since it isn't even necessary to have a mind at all in order to be a perfectly competent game player (although presumably it helps to have a mind in order to be a game *theorist*!). In any event, it seems likely that self-awareness is especially useful in a game-playing world, because it helps its possessors make good guesses as to how someone else is feeling or thinking, and, thus, how someone else will behave.

It is even possible that we owe not only consciousness, but even our vaunted rationality to game playing at its deepest levels. At issue in this case is not the payoff of solving the kinds of theoretical mind games we'll explore in the pages to come, but the real-world payoff, accumulated over eons, from being a winner.

In some cases, game theory actually helps advise participants what to do, but in fact during the half century or so that it has been around, the formal structure of game theory hasn't really given very much advice to average people trying to navigate the shoals of everyday life. Maybe someday it will; as we'll see, it already does for animals. As humans unravel its secrets, game theory has changed some aspects of our own behavior, notably in the realm of strategic, nuclear doctrine and the specific design strategies of some financial auctions. In an intriguing example of knowledge and its effects being recursive, knowing the rules of our games has begun to change the way we play. For now, however, it seems likely that game theory is most useful as a way to clarify our thinking, to see complex matters in a new way, and—if you enjoy occasional forays into logic and paradox—to have some fun with your gray matter.

One more thing: You'll find that for all the mathematical abracadabra (most of which we'll ignore) game theory is actually an oversimplification. But that's not necessarily a bad thing. After all, it is intended to be a model, and models by their nature—if they're good at their job—are simpler than their subjects. So *of course* the games we're going to discuss are simpler than life itself. If they weren't, then there wouldn't be any reason to think about them; instead, we'd just slog through the myriad details of every actual interaction, treating each one as something new and altogether unknown.

There you have it. That's the big idea. If it seems complicated and confusing right now, just be patient. By the time you've finished this book, it'll be obvious. And you'll wonder how you ever managed to play so many games, and to be such a canny strategic mathematician, without knowing it.

Sartre's Dictum

In Jean-Paul Sartre's play *No Exit* we are given this stunning line: "Hell is other people."

The great existential philosopher wasn't a misanthropist. Instead, he was a firm believer in freedom and the power—indeed, the obligation—of human beings to choose their own course of action. For Sartre, we are all "condemned to be free." Accordingly, it can be sheer hell when others get in the way. (It is said that when asked to account for the Confederacy's defeat in the Civil War, George Pickett—famed for Pickett's Charge, during the Battle of Gettysburg—replied: "I think the Yankees had something to do with it.")

For game theorists, other people aren't hellish; rather, they are the reason game theory exists, and why it is worth knowing. For the rest of us, there is nothing diabolical about the fact that other people exist, and that they have their interests, which often compete with, complement, or otherwise interact with our own. It contributes much—maybe all—of the spice of life. But at the same time, it complicates things, and immensely so.

Robinson Crusoe, alone on his desert island, is able to pursue a simple strategy: seeking the greatest good for the greatest number . . . an easy matter when the number is one. The Crusoe Course is simply to decide what yields the highest payoff, then go ahead and do it. Even if the payoff is not guaranteed, choose the one that, on average, is likely to yield the best outcome. Add someone else—in the Robinson Crusoe story, "his man Friday"—however, and things immediately become more complex. An engineer designing a building or a bridge need not worry that gravity or a river is plotting against him. Of course, gravity and hydrodynamics go their own way, and cannot be directly controlled by the engineer, but it is possible to account rather precisely for them. In simple games, the actions of each side are also uncontrollable (by the other player) but, worse than gravity or water, they *are* controlled by independent minds (minds that may well be as good or better than one's own), and, worse yet, they are likely to be aiming for results that are directly opposed to yours.

Furthermore, these minds are essentially secret and private, as hidden from ourselves as we are from them. According to Charles Dickens, in *A Tale of Two Cities,* it is

> a wonderful fact to reflect upon, that every human creature is constituted to be that profound secret and mystery to every other. A solemn consideration, when I enter a great city by night, that every one of those darkly clustered houses encloses its own secret; that every room in every one of them encloses its own secret; that every breathing heart in the hundreds of thousands of breasts there, is, in some of its imaginings, a secret to the heart nearest it! . . . In any of the burial-places of this city

through which I pass, is there a sleeper more inscrutable than its busy inhabitants are, in their innermost personality, to me, than I am to them?

And yet, despite all this secrecy and inscrutability, we have no choice but to interact with one another. Hell? Sometimes. Especially if you like your answers simple, your options straightforward and linear. And if you don't live alone on a desert island.

Most efforts to model and understand human behavior used—and still use—an approach derived from classical physics: Nongame theorists generally assume that objects (including people and animals) respond in certain predictable ways to the action of known external forces. Rarely incorporated into these models was the notion that these "external forces" may have their own agendas, and, furthermore, that the subject whose behavior is to be predicted and understood is likely to be acting with those others in mind. ("I know he knows. And I know that he knows that I know. And I know that he knows that I know that he knows . . .") And even game theorists focused largely on people, developing most of their models under the assumption that the players were conscious strategists.

Decision making can be complicated, even when there is only one player, and even when that player has complete access to all relevant information (what game theorists call "decision making under certainty"). For example, consider simple "optimization problems," in which the goal is to find the best way to accomplish something. For game theorists, such problems are not especially intriguing, since the idea is simply to find the most efficient way of achieving a particular goal: minimize time spent, maximize profit, construct the strongest or shortest link between two points, and so forth. In such cases, there is no other player in any meaningful sense, aside from nature. And often, nature does not respond.

Nonetheless, even so simple a decision as whether to take an umbrella when you leave the house involves several considerations: How awkward is it to carry? Are you likely to forget it somewhere? Does it make you look like a dork? But there is at least one thing that you don't have to factor into your decision: Carrying an umbrella doesn't make rain any more likely. Yet so-called optimization problems can still become fiendishly complicated. They are the stuff of the specialized mathematical discipline called linear programming (which is allied to game theory, but distinct).

Take one example, the Traveling Salesperson Problem. Imagine that a salesman must visit ten different cities, stopping in each no more than once. What order of visitations will produce the minimum traveling distance on his part? Even in this seemingly straightforward case, there are 3,628,800 different possibilities! Why so many? Our understandably bewildered salesman can start his tour in any of the ten cities, after which he can visit any of nine, followed by eight, and so on. The number of options is therefore $10 \times 9 \times 8 \times 7 \times 6 \times 5 \times 4 \times 3 \times 2 \times 1 = 3,628,800$. For twenty cities, the number leaps to 2,432,902,008,176,640,000. And for thirty cities it exceeds 265 thousand billion billion billion, far more than the number of subatomic particles in the visible universe![1] Take, then, the fifty state capitals in the United States and ask: How would you proceed if you wanted to visit all of them while flying the shortest possible distance? (Let's make the simplifying—and inaccurate—assumption that each state capital has its own airport and each one provides nonstop service to every other.) It turns out that in this "simple" case, it is literally impossible to make the best possible decision, even under conditions of certainty. This is not to say that the best possible route doesn't exist, just that there is no way to determine it.

Certainty is not the same as simplicity.

Even less simple, in a way, is the problem of making decisions under profound *uncertainty*, when there is another side and that other side is also making its own decisions. In some cases, these "decisions" aren't even conscious, and may even be counter-intuitive. Moreover, they may even involve players whose identity is unclear. Take dieting. This has long been seen as a simple matter of willpower (or won't power). But endocrinologists report that a consequence of self-starvation is that one's body responds as though there is a famine, thus reducing its metabolic rate. The result is that even something as apparently self-directed and inner-controlled as whether to eat less and, thus, lose weight turns out to be a case of playing a game against one's own body: You can control your eating (up to a point), but you can't control your body's response physiologically. And even if you and your body agree that you have a shared interest in losing weight, your brain and your fat cells may disagree.

As we shall see, it gets even trickier when the other side is more cognitively sophisticated than a glob of adipose tissue; in short, when the other side is taking the likelihood of your decision into account. Add to this the fact that in many cases each decision maker's interest may be diametrically opposed.

Most of the games we'll examine involve just two players, largely because adding additional participants makes things unwieldy, even for the mathematically adroit. At the same time, it is worth noting that sometimes a third party actually simplifies things or, rather, settles them down. Consider the following (somewhat) hypothetical game of geopolitical intrigue: Imagine that the newly installed government in Afghanistan isn't to Pakistan's liking. Pakistan may be tempted to invade Afghanistan and establish a more pliable regime; after all, Pakistan is much stronger militarily. So the situation could be dangerously unstable. But add a third player—India—and things become not simply more complex but, ironically, more stable. Thus, under this scenario Pakistan would likely inhibit its aggressive inclinations for fear that if it deployed large numbers of troops into Afghanistan, it might dangerously weaken itself vis-à-vis its border with India, which might then invade. So (even ignoring the role of the United States, world opinion, and numerous other factors) sometimes the addition of a third participant may stabilize interaction between two.

The more the stabiler? Not necessarily. Add a fourth player, China, and what happens? Given the historical distrust between China and India, the fact of China's existence would probably inhibit India in the (unlikely) event that Pakistan invades Afghanistan. This is because an Indian invasion of Pakistan might well weaken India vis-à-vis China, just as a hypothetical Pakistani invasion of Afghanistan could tempt India. In cases like these, the progression can be drawn out indefinitely, at least in theory: even numbers of players threaten instability, odd numbers promise the opposite. (I'm not familiar with parallels in the nonhuman world, but I can't help wondering whether this simple arithmetic might trickle into our biology and, thus, augur poorly for the stability of biological twosomes, notably monogamy.)

In other cases, game players needn't worry about just one individual on the other side; they have to consider what large numbers of others are doing. Let's imagine that you are about to purchase a new computer and, furthermore, that you have a preference for Apples. You have second thoughts, however, for the simple reason that PCs are much more popular, and, as a result, there is some risk that Apple might go belly-up, leaving you with a machine devoid of customer support and with very low trade-in value. A more immediate problem is this: because more people have PCs, there are more software options for them. On the other hand, if—as has been happening in recent years—Macs begin to make a comeback, the payoff to buying a Mac goes up. Success breeds success. Your payoff depends on what others do.

Buying a stock, especially in hope of making a short-term killing, would be another example. The question in this case is not simply whether the company being invested in is likely to make profits in the future. If so, this would be equivalent to choosing a computer based only on whether you like it, regardless of whether anyone else does. And investing in a particular stock would be like deciding whether or not to carry an umbrella, a decision based almost entirely on your assessment of whether it might rain, not on whether other people are going to carry their own umbrellas. Instead, short-term investors must also ask themselves whether their stock pick is likely to be attractive to *other* investors, and, therefore, whether its stock price is likely to go up or down as a result.

Famed economist John Maynard Keynes pointed this out in an oft-quoted passage:

> Professional investment may be likened to those newspaper competitions in which the competitors have to pick out the prettiest faces from a hundred photographs, the prize being awarded to the competitor which most nearly corresponds to the average preferences of the competitors as a whole; so that each competitor has to pick, not those faces which he himself finds prettiest, but those which he thinks likeliest to

catch the fancy of the other competitors, all of whom are looking at the problem from the same point of view. It is not a case of choosing those which, to the best of one's judgement, are really the prettiest, nor even those which average opinion genuinely thinks the prettiest. We have reached the third degree where we devote our intelligence to anticipating what average opinion expects the average opinion to be.[2]

Maybe we should update Sartre's epigram. Purgatory is other people. Hell is when we don't have a clue about them. And game theory—to damn it with faint praise—is at least a ticket from the latter to the former. In short, although it can be useful and even fun, it's no stairway to heaven.

Isn't It All Too Machiavellian?

It's one thing to use game theory to figure out an optimum strategy for betting at cards, managing a baseball team, or maybe even developing a competitive business plan, but quite another to be "game theoretic" when it comes to human interactions. (Hence, it seems more acceptable to be a game theoretician when playing stock market games than when engaged in a domestic "Battle of the Sexes," as we'll explore shortly.) There is something Machiavellian—cold, cynical, selfish, and calculating—about analyzing human situations and then basing one's decision on nothing more than this: What generates the highest payoff. My point, however, is that for better or worse, this is precisely what many of us—perhaps most—actually do. Like the centipede, who fell all over himself when asked to explain how he coordinates the movement of all his legs, people often feel intellectually or ethically paralyzed when asked how they arrive at competitive, interactive outcomes.

Not so with animals. They don't seek to justify their behavior in terms of its potential payoff gain (at least, so far as we can tell, they don't!). And yet, maximize their payoffs is precisely what they do. Given the choice between foraging in a food-rich environment and one that is depauperate, every animal that has ever been tested goes for the former. Machiavellian? Perhaps so, but few people—and probably no animals—would see this as reprehensible. Similarly, a sexually aroused male Tungara frog, just yearning to fill the night air with his ardent vocalizations, will nonetheless inhibit himself if, by cheerily croaking away, he would be the only one doing so; there are fringe-lipped, frog-eating bats that home in on croaking male Tungara frogs, so it is very much in the interest of even the horniest male amphibian to croak in a chorus or not at all. Hence, frog choruses, a Machiavellian strategy if ever there was one, whereby each male reduces his personal risk by spreading some of it to others.

The point is this: Strategies happen. It isn't a matter of basing decisions on Machiavellian payoffs, but of recognizing that we—like Tungara frogs—are doing so naturally. Indeed, the more skilled all of us (frogs and folks) become at seeing value in payoffs, strategic cooperation, and so forth, the better off we all may be. At the same time, the better we are at recognizing what others are doing, the better we can be at making good choices for ourselves.

Nonetheless, there is something in us that bristles at the idea. We experience ourselves as possessing—if nothing else—free will. Not surprisingly, therefore, most people resist formulations that seek to reduce our actions to simple cause-and-effect relationships. "Experience teaches us no less clearly than reason," writes Spinoza, "that men believe themselves to be free, simply because they are conscious of their actions, and unconscious of the causes whereby these actions are determined." Most people are especially resistant, in fact, to the suggestion that their actions are "determined" by anything so crass as maximizing their "payoffs."

At least part of the problem is that human beings generally cherish an image of themselves as kind, generous, inclined to "do the right thing," and they attribute that inclination to nothing other than the fact that it *is* the right thing. If a person does something—even the most generous act—for an identified "payoff," then by definition it no longer appears generous. We are suspicious of the "altruist" who admits that he is seeking social approval, or even divine ratification, or attempting to induce future reciprocation by the beneficiary. The true, admirable altruist (quotation marks removed) is one whose actions are supposed to be uncaused, or, better yet, caused only by purity of heart.

So let's turn directly to that most Machiavellian of thinkers, Niccolò Machiavelli himself, who wrote: "A man who wishes to make a profession of goodness in everything must necessarily come to grief among so many who are not good." And also:

Men must either be caressed or else annihilated; they will revenge themselves for small injuries, but cannot do so for great ones; the injury therefore that we do to a man must be such that we need not fear his vengeance.

And finally, here is Machiavelli's famous advice on whether it is better for a ruler to be loved or feared:

One ought to be both feared and loved, but as it is difficult for the two to go together, it is much safer to be feared than loved, if one of the two has to be wanting. For it may be said of men in general that they are ungrateful, voluble, dissemblers, anxious to avoid danger, and covetous of gain; as long as you benefit them, they are entirely yours; they offer you their blood, their goods, their life, and their children . . . and men have less scruple in offending one who makes himself loved than one who makes himself feared; for love is held by a chain of obligation which, men being selfish, is broken whenever it serves their purpose; but fear is maintained by a dread of punishment which never fails.

Still, Machiavelli argued that

a prince should make himself feared in such a way that if he does not gain love, he at any rate avoids hatred; for fear and the absence of hatred may well go together, and will be always attained by one who abstains from interfering with the property of his citizens and subjects or with their women. And when he is obliged to take the life of any one, let him do so when there is proper justification and manifest reason for it; but above all he must abstain from taking the property of others, for men forget more easily the death of their father than the loss of their patrimony.

The reason for quoting Machiavelli at length is not simply to call attention to his up-front cynicism, almost refreshing in its brutal honesty, but to point out his game theory relevance (which is not simply its "up-front cynicism, almost refreshing in its brutal honesty"!). Rather, it is because Machiavelli, for all his "unprincipled" recommendations, reflects a basic game theory principle: *We must consider not only our interests but also the interests and behavior of others,* and how those others are likely to respond to our actions. Thus, a prince, according to Machiavelli, should be scrupulous with regard to his subjects' property rights because those subjects are likely to take special umbrage if those rights are infringed. Because love and loyalty are fickle, a prince should rely, if need be, on fear. Again and again we find in Machiavelli that sound strategy depends on a careful assessment of what *others*—one's subjects or a foreign leader—are likely to do.

Such thinking isn't inherently right-wing, or militaristic. One could just as well offer comparably Machiavellian, game-theoretic political advice that is progressive, left-leaning, and pro-peace. It's really a matter of seeing the degree to which "selfish" benefit might be achieved by taking the other side into account. For every argument that equates self-interest with going it alone, there is another that emphasizes the benefits of cooperation. Taking the other side's perspective into account is not, in itself, compatible with either right-wing or left-wing political theory, but is very definitely the way of game theory.

Here's yet another, more general way of looking at it, from negotiation experts Roger Fisher and William Ury, no Machiavellians, they:

> The ability to see the situation as the other side sees it, as difficult as it may be, is one of the most important skills a negotiator can possess. It is not enough to know that they see things differently. If you want to influence them, you also need to understand empathetically the power of their point of view and to feel the emotional force with which they believe it. It is not enough to study them like beetles under a microscope; you need to know what it feels like to be a beetle. To accomplish this task you should be prepared to withhold judgement for a while as you "try on" their views. They may well believe that their views are

"right" as strongly as you believe that yours are. You may see the glass as half full of cool water. Your spouse may see a dirty, half-empty glass about to cause a ring on the mahogany finish.[3]

The Regrettable Reality of Conflict

Game theory is a theory of conflict. Fortunately, it also offers a powerful theory of cooperation, which we shall get to. But first, this hard truth: If mutual agreement were at the root of most interactions, there would be little for game theoreticians to theorize about. As it is, there's quite a lot.

So long as there are different individuals, there are going to be conflicts of interest, not only among those individuals but also among the groups formed by them. Let's look at a few relevant words.

The word *rivalry* originated with the Latin *rivus* (river or stream). Rivals were literally "those who use a stream in common." Competitors, by contrast, are those who compete (from the Latin *com* meaning "together," and *petere*, "to strive"). Competitors strive against one another by seeking to obtain something that is present in limited supply, such as water, food, mates, or status. Rivals necessarily compete, whenever a sought-after resource is present in limited supply. This is unavoidable. But they do not have to be enemies.

The word *enemy* also derives from the Latin, this time *in* ("not") plus *amicus* ("friendly"), and it implies a state of active hostility. The word *conflict,* on the other hand, derives from *confligere*, which means literally "to strike together." It is impossible for two physical objects—like two billiard balls—to occupy the same space. They conflict, and if either is in motion, their conflict will be resolved by a new position for both of them. Within the biological realm—whether human or animal—conflict occurs when different individuals or social entities are rivals or otherwise in competition. Such conflicts can have many different outcomes: one side changed, one side eliminated, both sides changed, neither side changed, or (rarely) both sides eliminated. Conflicts can be resolved in many ways: by violence, by mutual agreement, by issues changing over time, and so forth.

One possibility, and in some ways the simplest means of overcoming enmity, is to diminish competition itself. South of Worcester, Massachusetts, near the Connecticut border, there is a small lake with the wonderful Mohican name of Chargoggagoggmanchauggauggagoggchaubunagungamaugg. In English, it means "You fish your side, I fish my side, nobody fishes the middle: no trouble." We may assume that the early lakeshore residents of this mellifluously named body of water were not enemies. They also knew something about managing human affairs. By diminishing competition—thereby, in a sense, overcoming it—they overcame enmity. But what if, at some time in the future, the inhabitants of one side started to cheat? And what if, anticipating such a possibility, the inhabitants of the other sought to do so first?

At this point, an indignant reader might point out that such perverse questioning is itself part of the problem. If people didn't worry about the cheating of others, and hurry to preempt them, much misery could be avoided. By the same token, game theory can be accused of potentially provoking conflict, because it anticipates it, thereby threatening to become a self-fulfilling prophecy. The idea is that in some cases, predicting certain behavior can contribute to its taking place: anticipate the enmity of the other side, and behave accordingly, and sure enough, you've made an enemy. National security analysts have even coined a term to describe this self-fulfilling prophecy: *the security dilemma*. It arises when a country attempts to enhance its security by increasing its military power, which in turn reduces the security of its neighbors, who respond by increasing *their* military power . . . which leaves the first country—and everyone else—less secure than before.

Game theoreticians have been accused, sometimes correctly, of contributing to precisely this perverse outcome. But at the same time, there are many potential ways out, notably by using game theory itself to anticipate the worst so as not to be blindsided, and also to avoid provoking the result that one wishes to avoid. It is one thing to be prepared, quite another to cause the problem you dread. A special strength of game theory should be that it is explicitly *not* one-sided; it incorporates concern for the other players, and recognition of their interests, if only because this leads to a more realistic prediction of outcomes for all concerned.

At the same time, an alternative perspective also deserves hearing. Even as the danger looms that anticipating conflict could result in precipitating it, there is also a hidden, potential risk in denying its likelihood. Here's how it works. Peace is generally thought to be the absence of conflict. Moreover, peace is often considered to be more natural than conflict, perhaps because it is more desirable. Fair enough . . . except that the results of this perspective can be troubling in their own right. If you truly believe that the normal state is for the lion to lie down with the lamb, for people to live together in unconflicted bliss, then you are likely to feel especially annoyed when difficulties arise. As a result, when conflicts of interest emerge—as they inevitably do—well-meaning but disappointed idealists are sorely tempted to blame someone for upsetting the peaceful applecart. Convinced that serious evil is afoot, the next step may be to eradicate the evildoer.

By contrast, a more realistic expectation—namely, that conflict, like shit, happens—may lead to less outrage and, in the end, a less violent resolution. Game theory is built upon the idea that conflict is highly likely if not inevitable, but that it needn't necessarily be violent or destructive. Marriage counselors and family therapists, similarly, rarely urge their clients to avoid conflict and to seek, instead, a situation of uninterrupted harmonious bliss. Rather, the goal is to learn how to *manage* conflict effectively, creatively, and nonviolently. When stuck with lemons, make lemonade.

Here's a simple case of conflict, readily resolved, with a little help from game theory. Two people want to divide a piece of pie. The simple solution—remarkably sophisticated in its assumptions—is for one to cut and the other to choose. Neither person is presumed to be especially nasty or otherwise disreputable; nonetheless, each wants as much as possible. But since Cut the Pie is a "zero-sum" game, each player is aware that whatever the other gets means that much less for oneself. Also, neither can influence the behavior of the other—at least, not directly. Each assumes that the other will pick the largest possible piece. A simple solution might be for the cutter to divide the pie so that there is a very large piece and a very small piece, hoping that the chooser will take the small one! But obviously, this would be a terribly naïve assumption, the kind of wishful thinking that results from a failure to take into account the fact that others are as capable of enlightened self-interest as is oneself. Instead, the rational cutter assumes that the chooser will try to maximize her piece, and so, he tries to make the two pieces as similar as possible. The solution to this game is a 50:50 division, or something as close to 50:50 as a human being—armed with a knife, rational self-interest, and a realistic perception of another's likely behavior—is able to get.*

The result is an equilibrium at which each player does the best he can, given his knowledge that the other is doing the same.

Here is another way of looking at it. An important part of game theory recognizes that, paradoxically, even conflict tends to settle down. Or in mathematical language, it reaches an equilibrium. What kind of equilibrium? Often, one at which each player is making the best possible response to the other, who, in turn, is making his or her best response to the first. (This is a so-called Nash equilibrium, named for mathematician John Nash, winner of the 1994 Nobel Prize in economics; sadly, however, even though Nash equilibria are stable, they don't always produce optimum outcomes for the participants; perhaps the best examples come from the renowned—or notorious—game of Prisoner's Dilemma, discussed in chapter 3.)

For a comparatively benign example of how players can settle, unconsciously and at least somewhat peacefully, into equilibrium strategies, take this case, described by Stephen Jay Gould.[4] Gould didn't intend it to illuminate game theoretic equilibria, but it does. He was interested in something more mundane: baseball. More specifically, he was intrigued with the fact that batting averages in baseball consistently declined during the twentieth century. Although there were never very many .400 hitters, for example, there were far more such "phenoms" early in baseball history than in recent times. One explanation: There simply were better athletes in the good old days. But wait: Why shouldn't there also have been better pitchers, which would have kept the batting averages in check?

---

*Actually, even pie cutting is not as simple as one might think! See *Fair Division* by Steven Brams and Alan Taylor (Cambridge University Press, 1996), as well as Brams and Taylor's *The Win-Win Solution* (Norton, 1999).

Analyzing decades of baseball statistics (baseball yields nothing if not statistics), Gould demonstrated that not only are there fewer exceptionally good hitters in modern times, there are also fewer exceptionally *bad* ones. To be sure, there aren't many .400 hitters these days, but neither are there many who hit less than .200, or even .220. Gould made the point that modernity as such has produced a reduction in the extremes, as pitchers and batters have gradually adjusted to each other:

> When baseball was very young, styles of play had not become sufficiently regular to foil the antics of the very best. Wee Willie Keeler could "hit 'em where they ain't" (and compile an average of .432 in 1897) because fielders didn't yet know where they should be. Slowly, players moved toward optimal methods of positioning, fielding, pitching, and batting—and variation inevitably declined. The best now meet an opposition too finely honed to its own perfection to permit the extremes of achievement that characterized a more casual age.

We can have confidence in Gould's explanation because this system—in which offensive and defensive players gradually develop an equilibrium of mutual best responses to each other—tends to break down whenever the rules are changed. Things were thrown out of equilibrium, and variation increased (that is, more high and low batting averages) when the pitching mound was lowered, when new teams entered, when the size of the strike zone was changed, and even when new technology was introduced, such as a livelier ball or different kinds of gloves. In such cases, the balance was typically shaken for a time, during which offensive and defensive players experimented with different ways of doing things; some did unusually well, others were exceptionally inept. Eventually, a new equilibrium was typically established, with each not only doing the best he could, but—crucial for our game theory perspective—the best he could *against the other.* Each side's "best" depended (and continues to depend) on what the other is doing.

Baseball is only a game. But so are many other things.

What It'll Tell Us and What It Won't

A criticism sometimes raised is that game theory is too encased in its own logical analysis; it doesn't tell us anything that isn't already present in the underlying assumptions of the system. Like 2 + 2 = 4. (I know: I promised no equations, but this one hardly counts, nor do the others appearing just ahead.) Two plus two *must* equal four, because of how we define two and four, the latter being twice two. But, in fact, self-contained systems can be very interesting, or at least, they can include a lot of richness not immediately apparent. For example, take the Pythagorean theorem. In a right triangle, the hypotenuse squared equals the sum of the square of the other two sides. This is necessarily true, given the nature of right triangles. It flows automatically and ineluctably from the definition of this particular kind of geometric shape. Yet it isn't immediately obvious. It had to be worked out by a mathematical genius. Even though the relationship of hypotenuse to sides is embedded in the very definition of "right triangle," it wasn't known until Pythagoras pointed it out. Since then, it has served as the cornerstone for much of geometry, most of which is also logically self-contained (as is algebra, by the way).

Thus, we can solve a simple equation (say, 3x + 4 = 10) by performing certain operations (subtracting 4 from each side, then dividing by 3), only because in doing so, we aren't really changing the equation in any way. By performing these simple operations, we reveal something *already contained* in the equation, the fact that x = 2. But that doesn't mean that we haven't revealed something important. (Or at least, something that many struggling algebra students don't consider self-evident.)

More than this, in solving this equation we have used formal logic to tell us something that most of us, at least, could not immediately discern. That's what game theory does: It tells us things that are true but not immediately obvious. In the case of algebraic equations, the "value" of x varies with the initial conditions: 5x + 6 = 21 specifies a different value of x than does 3x + 4 = 10. Simple algebra enables us to reveal x once any particular condition—a specific equation—is given. Game theory will not tell us anything about the world in general, any more than algebra will tell us the value of x in general, but it will allow us to learn a whole lot once the particular starting conditions—the nature of the game—is made clear.

By stripping away extraneous material, game theory can help us look clearly at what's at stake in many situations, especially those involving cooperation or competition. At the same time, we have already noted that game theory oversimplifies reality. In most of the cases we'll be examining, each player will have two choices (either go straight or swerve, either accumulate weapons or don't, either look for new sexual partners or stay home, and so forth), whereas, in fact, individuals as well as organizations usually have more than two options from which to choose, no matter what is going on. But for simplicity's sake—or, as mathematicians put it, to make the analysis "tractable"— it is often necessary to reduce each player's array of options to just two. Supporters of game theory point out, however, that the basic principles don't change no matter how many courses of action you consider; relying on simple models to help us understand complex reality is essentially a matter of walking before we can run. And devotees of game theory have reason to believe that they've at least been getting somewhere.

Nonetheless, here is a bit more complexity that must be taken into account: Even when payoffs depend on another's action, in most cases, people aren't helplessly dependent on what the other party does. They can talk to each other, for example, and try to persuade, cajole, threaten, or point to the better angels of human nature: one's own, if attempting to convince someone else that you are going to "do the right thing," or someone else's, if attempting to convince that someone else to behave angelically. Supporters of game theory point out, in turn, that no matter what anyone may say, what really counts is the moment of truth when someone actually acts (as distinct from talks), and that a rich texture of discussion doesn't negate the underlying reality of payoffs; it just adds to the information that any rational player must consider in deciding on the next move.

That's not all. Critics sometimes note that game theory, with its dependence on identified payoffs and its assumption that players are merely trying to obtain the highest payoff possible, fails to give other factors their due: For example, some players may be more altruistic and less selfish than game theory assumes. They may seek a different, and perhaps even a higher reward, such as the pleasure of helping others (doing good being its own reward), or perhaps greasing their way into heaven. To this, game theoreticians respond that such considerations, even if true, don't invalidate the notion that different actions produce different payoffs, or that people seek to obtain the highest payoff possible; it is simply necessary to consider how people value the consequences of their behavior, which might require changing the payoffs received by certain people for certain actions. This is very different from denying the existence or relevance of payoffs themselves. For a devoutly religious Christian, turning the other cheek might have a very high payoff value. For someone else, it may merely result in a painful slap in the face.

Game theory does not necessarily assume personal selfishness. People can, for example, prefer *not* to receive a particular payoff, in which case it takes on a negative value. Or perhaps they would like to donate a particular payoff to charity . . . in which case it still has a positive value to them, but because of its "give-away value."

Contrary to what some might think, game theory is not necessarily predisposed toward those who are especially Machiavellian, amoral, or immoral. There is no reason why a payoff couldn't take all sorts of highly ethical considerations into account. There is not even a presupposition that winning is always the goal. An adult playing checkers or tic-tac-toe with a child, for example, might ascribe a high payoff to *losing*. For a dyed-in-the-wool altruist, there could presumably be a high payoff to helping someone else, even at substantial personal cost. This simply means that for such an individual, a different payoff value must be assumed, not that there is no possible set of rewards that accurately reflects his or her preferences. Reformulating one's payoffs to take this into account doesn't show the power of "true altruism." Quite the contrary: It nearly always italicizes the self-aggrandizing nature of the underlying motivation. It suggests that even the most self-denying saint is deriving personal—dare we say, selfish?—benefit from her actions.

Nor is this pattern limited to human beings. For example, one of the most important discoveries in evolutionary biology concerns the nature of animal altruism. It is now widely acknowledged that an act that appears to be self-abnegating—benefiting another at the expense of the "altruist"—can actually be selfish at the level of individual genes, insofar as copies of the genes in question are present in the beneficiary. Since genetic relatives are, by definition, individuals likely to share genes, altruism directed toward them warrants a payoff value that is high rather than low. Thus, when a prairie dog "barks" an alarm call, thereby warning its kin of an impending attack by a coyote, the alarmist—more properly, genes that lead to alarm calling—may well be obtaining a substantial positive return on his or her behavior, since within the bodies of those prairie dogs thereby alerted there will reside copies of alarm-calling genes, which profit from the alarm calling itself. Altruism at the level of bodies (negative payoff) is typically selfishness at the level of genes (positive payoff).[5]

Maybe the unpleasant connotations of game theory do not reside in the models themselves so much as in the very idea of externalizing personal motivations. There is something hard-eyed and seemingly amoral in the very effort to identify and—worse yet—to quantify motivations, even if one gives a very high payoff to "kindness," "ethical behavior," or "doing what is right, simply because it is right." Face it: We don't like being told that there are reasons why we do things. The extent to which people take—or give—credit for an action seems to vary inversely with the degree to which that action can be "explained." Somehow, a rational basis for behavior seems to diminish our individuality, our free will, or our goodness. It is as though many of us would rather believe that our actions are uncaused, especially those behaviors in which we take particular ethical pride. Good actions are supposed to flow of their own accord, rather than be caused, certainly not caused by some sort of scheming, manipulative assessment, even if entirely unconscious, of alternative costs and benefits. And even if "doing the right thing" is valued for its own sake.

Another criticism is that game theory applied to behavior takes ends as given and concerns itself only with means to achieve these ends. In this regard, it may be no different from reason itself. Thus rationality may best be understood as a logical technique, the most effective way of gaining one's ends—what Max Weber called *Zweckrationalität* (instrumental rationality), as opposed to *Wertrationalität* (rationality of rightness). The former illuminates the most efficient way of going about achieving one's goals, whereas the latter is concerned with what these goals ought to be. Game theory is totally neutral concerning *Wertrationalität* (what an individual ought to prefer), although it is quite clear on *Zweckrationalität* (how to go about achieving a goal once it has been identified).

For Bertrand Russell, justly renowned for his contribution to both mathematical rationality and ethical analysis, "reason has a perfectly clear and precise meaning. It signifies the choice of the right means to an end that you wish to achieve."[6] (Professor Andrew Colman has pointed out to me that this insight was prefigured by the eighteenth-century philosopher David Hume, in his *Treatise of Human Nature*: "A passion can never, in any sense, be called unreasonable, but when founded on a false supposition, or when it chooses means insufficient for the designed end.")

For game theorists, rationality means acting so as to bring about the most preferred possible outcome, given that other players are also trying to achieve the same ends. It does not speak to the question of how those outcomes are ranked: It may be perfectly rational for someone to prefer beer over wine, and for someone else to feel exactly the opposite. But if you do prefer beer, then it is irrational to buy wine . . . unless you are buying it for someone else, or for some other purpose.

Game theory never questions the rationality of the utilities (values or goals) people employ. It simply asks: Given these goals, and their rankings, what is the most effective way to achieve them? It assumes that the players are rational, but this simply means that they choose from among the available options the one most likely to achieve their goals, what they value most. It doesn't concern itself with *what* people value, or why. So, if someone wants to lose a game, then that's his highest payoff. If he wants to suffer pain, that's his choice, even if others might think such a goal "irrational." The issue then becomes: Given that someone wants to suffer pain of a particular sort, what is the most effective way to achieve it? Similarly if the goal is to inflict pain, to win the largest amount of money, and so forth.

At the same time, game theory carefully applied can help us find out what people are valuing, by helping us focus on what payoffs they are obtaining, or maneuvering toward. (This assumes, of course, that they are rational in that at least to some extent their strategies are oriented toward obtaining whatever it is that they want.) In evolutionary studies, for example, it is increasingly clear that living things seek to maximize their "fitness"—their success in projecting genes into future generations—and this clarity has been achieved, in part, by noting the degree to which animal behavior accords with the predictions of game theory whenever reproductive success is at stake. By contrast, if biologists were to assume that animals are maximizing their oxygen intake, or their time spent sleeping, or their probability of bumping into one another, such predictions would not be supported. Game theory thus helps us understand what living things are up to.

You might want to think of game theory as being like National Public Radio (NPR), or a microwave oven: neither is essential, yet both are useful. Game theory can be an enhancement to our understanding (like NPR) and in some cases, a convenience (like a microwave oven). I don't recommend relying on NPR, and nothing else, for news, or cooking only with microwaves, but at the same time, I wouldn't ignore the former just because *The New York Times* or the local television news does a better job with certain things, or discard my microwave oven because grilling is far preferable when it comes to cooking a steak.

Later, we'll take a look at rationality itself, and ask to what extent people actually behave rationally, however defined. But for now, let's leave it that game theory seems to be a useful tool, not an end in itself. It can help clarify our thinking and, in some cases, even enable us to have a rollicking good time playing with our own minds and seeing familiar things in new ways. It isn't, however, a holy grail, or even a map for how to find it.

Duels, Truels, and Rules

Game theory violates one of the most beloved American myths: the rugged individualist, the notion that success comes to those who go out on their own and grab it by the scruff of the neck, wrenching wealth and happiness from the world, without regard to what anyone else does, or cares. Think of Arthur Miller's play *Death of a Salesman,* in which we learn briefly of Willy Loman's brother Ben, who went off to the wilds of Africa: "When I walked into the jungle, I was seventeen. When I walked out I was twenty-one. And, by God, I was rich!" It didn't matter whether there were any native Africans, whether Ben had any partners or victims. Ben Loman "made it," and he did so all by himself. It is a useful—if inaccurate—image for a growing country, expanding into a rich continent. Think of the rugged frontiersmen (and -women), carving a nation out of the wilderness, reveling in liberty and pursuing happiness. Or of the gunslinger, or the Wild West marshal.

But even here a kind of game intervenes. Call it the Gunslinger Game. Two gunfighters walk slowly toward each other in the dusty street. The outcome for each clearly depends not only on what he does, but on the other's action, too. Neither wants to get shot; each wants to shoot the other. There is a payoff to drawing your gun first and shooting your opponent, but since each seeks that same payoff, the situation is unstable, to put it mildly. A frequent outcome: Both draw and each shoots the other.

The game gets more interesting—and no less lethal—if we add this complication: Each still wants to shoot the other, but neither wants to be the one who draws first, because the initiator can legally be seen as the attacker. The problem, of course, is that the attacker is also more likely to be the victor . . . unless the responder is so fast that he can wait until the attacker attacks, then still fire first. In the real world, most gunfighters are not that confident. Neither are countries. (Animals, interestingly, often are. Thus, conflict among the bitterest of rivals typically involves lots of bluff and bluster, but little lethality. Even male rattlesnakes, capable of killing each other by a single bite, instead seek to push each other over; they keep their revolvers holstered.)

But human beings have discovered how to arm themselves via technology, and, as a result, they confront each other with weaponry that goes beyond the "merely biological." And as a further result, situations of this kind, in which each side fears being preempted by the other, are dangerously unstable. Think of the United States and the former Soviet Union during the darkest days of the Cold War. A great fear was that either side (or both) might reason as follows: "I'm not planning to attack, but they might be (or, they evidently think—incorrectly—that we are planning to attack them). As a result, they might well be intending to preempt us by attacking first. So, I better preempt their preemption." In the fateful days leading up to World War I, both the Allies and the Central Powers knew that whoever mobilized first would have an advantage. Germany feared that France and Russia would get the jump; France and Russia had similar anxieties. No wonder there was a war.

We'll see shortly that a duel, whether between people or countries, can be what is known as a Prisoner's Dilemma, in which the participants become locked into a devilish paradox whose outcome is disadvantageous to both. Dueling gunslingers, like countries playing brinkmanship, can thus find themselves pushed into shooting each other.

An interesting variant on the duel was first introduced by economist Martin Shubik,[7] and then elaborated by NYU political scientist Steven Brams.[8] Call it a truel, since it is a duel for three. Let's imagine three gunslingers, approaching one another equally, like three angles of an equilateral triangle. Assume further that they each have only one bullet (this isn't strictly necessary, but makes it easier to analyze). If the trio—call them Moe, Larry, and Curly—are all equally good shots, we might assume that, for example, Moe might shoot at Larry, Larry at Curly, while Curly fires at Moe. Instead, let's also suppose that one of them, Curly, is particularly incompetent. You might then think that he would be a dead duck, but, in fact, the ideal strategy for him—and, ironically, for the others—suggests that he has a fighting chance . . . in fact, the best chance of all.

Curly's ideal strategy is to wait. The two best shots should proceed to fire at each other, since each is the greatest threat to the other. Having refrained, Curly, the worst shot, can then fire with impunity at the survivor. In this way, the poorest shot is paradoxically assured the greatest probability of survival. (For a modern-day example, think of a hotly contested presidential primary election, in which the front-runner is prone to being knocked out early, since he is the target of everyone else's "best shot." In such cases, it may well be smartest to stay in the middle of the pack and not make your move until the leaders have damaged one another.)

A kind of truel has even been observed among animals, including certain fish. In these cases, large aggressive males—equivalent to hot-shot gunslingers—fight it out with one another while small, unobtrusive males take advantage of their mutual preoccupation to sneak in and attempt to fertilize the females.

In an actual human truel with guns, each individual has an even better and more paradoxical option: fire into the air. A player who did so would no longer be a threat to the other two; those two would have no reason to shoot at the disarmed individual and would be more likely, instead, to aim for each other. In fact, given the odds for survival, a disarmament competition might even ensue, in which all parties strive to be the first to waste their bullet, thereby removing themselves as a threat to the others! Alternatively, each ought to refrain from shooting at all, although *Reservoir Dogs,* directed by Quentin Tarantino, suggests otherwise: in the movie's climactic scene, three criminals confront one another in a tense triangle of drawn weapons . . . and everyone shoots everyone. Maybe it's just Hollywood. Or maybe more people should study game theory.

2

Mastering the Matrix

Life is one thing, filled with sneaky fish, adulterous bluebirds, murderous gunslingers, tough-talking politicians, nuclear armed rivals, and even (on a milder but more familiar level) interrupted telephone calls. Theory is another, filled with ideas and concepts. Yet the two are supposed to meet, with the latter—including game theory—helping us generalize about the former, in order to understand it better. When it comes to game theory, we can do surprisingly well without the mathematics, but to capture the vividness of life—at a glance—we need a special kind of picture.

Each picture is supposed to be worth a thousand words. In the world of games, it's probably worth a bit less. This is because the "pictures" used in game theory aren't like photographs or paintings; they're more like simple diagrams. Each one is known as a "payoff matrix," and, like a picture, it provides a quick snapshot, something that allows us to take in a lot of information in a moment or two. Since this is generally a worthwhile project, we'll spend the present chapter introducing basic games and their instantaneous depiction, the matrix.

By one definition, a matrix is "that which gives origin or form to a thing," or "the basic substance in which particular items are embedded." As a popular movie, *The Matrix* told of a fictional world in which a hideous reality was embedded in a virtual "matrix" of misleading appearances. Game theoreticians, too, employ their own version of a matrix, once again as a representation of reality, but one considerably more benign, more accurate, and, moreover, oriented toward understanding rather than deception. The word originates from the Latin *mater*, or mother, which may well overstate its importance in the present circumstance. There is no evidence, moreover, that the matrix loves us, or even that we should love it. Indeed, it should soon become apparent that unlike a mother, a matrix is not essential, although it is at least useful. And certainly, worth getting to know.

Coordination Games

The simplest games call for two players, each having just two possible moves. A good way to picture this is with a "payoff matrix," which is basically a 2 x 2 square, showing all possible combinations, along with the returns (the payoffs) to each player. Let's take the Interrupted Telephone Call Game, mentioned earlier. You're speaking with someone, perhaps long distance, and suddenly the call is interrupted and you get a dial tone. What to do?

Here are the options, for both of you, with the payoffs indicated for just one player, you:

| | | The Other Party | |
		calls	waits
You	call	−	+
	wait	+	−

If both talkers attempt to call back and bump into each other, you lose (wasting time and accumulating frustration), indicated by a "−" in the upper-left corner. If both of you wait for the phone to ring and it doesn't—because each is waiting for the other—you lose again, indicated by another "−" in the lower right. If you both somehow manage to coordinate things so that you call and the other waits, you reconnect, resulting in a positive payoff ("+"), shown in the upper right. Similarly, if you wait and the other party calls you, you receive another positive payoff ("+"), in the lower left. (It could be argued that this payoff—when you wait and the other party calls—is a bit higher than its mirror image, since in this case, you save the cost of the call, but we won't worry about that.)

In this example, because of the nature of the game, you and the other party get the same payoff in each case. The game is symmetrical. For most other games, however, this isn't true, so in the interest of mastering the matrix, let's include the payoff to both players in one simple diagram:

		The Other Party	
		calls	waits
You	call	− , −	+ , +
	wait	+ , +	− , −

Here, each corner shows your payoff first, followed by a comma, then the payoff for the other party. This is the general form of any payoff matrix: The return to the player on the left (sometimes called the "row" player) is presented first, then the return to the one on the top (the "column" player). You can see that there are lots of possible combinations, even if payoffs are merely represented as + or −. In fact, sociologist James Schellenberg has identified 881 distinct "primitive games," all involving this kind of simple 2×2 matrix, when the payoffs to each party are simply represented as either +, 0, or −.[1]

We'll try to keep things simpler yet, and look mostly at games for which there are only two possible payoffs to each player: + or −.

Before going on, however, let's look at a payoff matrix for another game we've already mentioned: the Gift of the Magi. In this O. Henry story, the husband sells his watch to buy his wife a comb for her exceptionally beautiful hair, while she sells her hair to buy a chain for his heirloom watch. Their payoff matrix is nearly the same as for the interrupted phone call, the only difference being that if both watch and hair remain unsold, there is no payoff (positive or negative) to either party:

		Wife	
		sells hair	keeps hair
Husband	sells watch	− , −	+ , +
	keeps watch	+ , +	0 , 0

To be sure, there are other considerations involved. O. Henry's story is lovely and bittersweet, not simply because of its paradoxical outcome, but also because the misunderstanding and mutual sacrifice bespeak the couple's love and commitment. Hair, after all, grows back, and the husband—described as hardworking, intelligent, and ambitious—can be expected to earn enough money some day to redeem his watch. In the meanwhile, their predicament (shown in the upper-left corner of the matrix as a mutually negative payoff) might not be so negative at all, if the couple basks in the warm glow of their acknowledged mutual generosity and devotion, perhaps taking a somewhat perverse pleasure in their shared plight, which, after all, italicizes their love.

At the same time, it is instructive—and, I hope, entertaining—to see this famous interaction in a new way, through the lens of game theory. (Like most lenses, game theory distorts in some respects and clarifies in others.)

The Gift of the Magi is a fundamentally benevolent game; it ends up with a negative payoff for both players because of a lack of coordination between them, not a lack of common interests. This and the Interrupted Telephone Call are "coordination games," because the participants have shared interest, with yet another mutuality added on: a positive, mutual payoff if they can both recognize their common plight/opportunity, and somehow coordinate their actions accordingly. In such cases, secrecy is inimical: It may be romantic to surprise your beloved with a new watch chain, but problematic if by keeping your intent secret, he winds up selling his watch to buy combs for you to use in your stunning hair . . . which you sold to buy the chain.

Some other coordination games: In the United States, people drive on the right side of the road. If you elect to drive on the left while pretty much everyone else is on the right, you are likely to be in big trouble. There is no inherent benefit to right-side driving over left-side; in fact, the coordination rule in England is to drive on the left. The important thing is that everyone agrees, for everyone's benefit. And no secrets.

Here is a simple matrix for the Driving Game. The American system is in the upper left quadrant; the British, in the lower right. The other two are anarchic, and very dangerous. Coordination pays.

		Jones drives on the	
		right	left
Smith drives on the	right	+ , +	− , −
	left	− , −	+ , +

In the Driving Game, the best payoffs arise when both players do the same thing (in this case, when they follow the "rules of the road"). When in London, do as the Londoners do. Ditto for Chicago. By contrast, in the Interrupted Telephone Call Game, the two players get their highest payoff when they do *different* things: One phones back while the other waits. In both cases, however, coordination is key, and the matrix helps us see the whole business, in a flash.

Coordination is not only important, but also—ironically—often taken for granted. Moreover, one of the rarely acknowledged advantages of honesty is that it facilitates the sort of coordination game that generates a solid alibi. Consider this example, widely distributed on the collegiate Internet, and probably (but not necessarily) apocryphal. Two students spent the night before a final exam drinking beer instead of studying. The next day, they were in no shape to take the exam but were embarrassed to explain why, so they agreed to tell the following cock-and-bull story to their professor: They claimed to have been driving a mutual friend to the hospital, whereupon they suffered a flat tire on the way home, and, not having a spare, were forced to stay up all night getting back to their dorm. The professor obligingly agreed to postpone the exam until the next day. The students studied hard—at last—and prepared adequately for the test. They were placed in two different rooms and handed the same make-up exam. The first question, worth 10 percent of the grade, was easy. The students were feeling pretty confident when they turned the page and encountered the second question, worth 90 percent. It consisted of just two words: "Which tire?"

As it turns out, our collegiate miscreants might not be totally lost, even at this point. For one thing, there is a one in four prospect that the two would, by chance alone, arrive independently at the same guess: assuming that each tire is equally likely to be chosen, there is a one in four probability that each student will settle on any one, and therefore the chance that both would make the *same* guess (e.g., left rear) is $\frac{1}{4}$ times $\frac{1}{4} = \frac{1}{16}$. But since any of four possible "same guesses" would do, there is, overall, a $\frac{1}{4}$ chance ($\frac{1}{16}$ times 4) that their guesses would correspond.

It turns out, however, that their prospects are better yet, since about 50 percent of people, when given the option, identify the same tire: front, driver's side. Although there is no obvious reason why this is so, our lives may well be filled with such unconscious focal points, helping us achieve beneficial, coordinated outcomes without even trying. In a survey conducted several decades ago, people were asked where they would go to meet someone "in New York City." A high percentage responded with "Times Square," "Grand Central Station," "Penn Station" and "The Empire State Building." One highly educated woman suggested the reading room of the New York Public Library; when told that not many others shared her choice, she pointed out that this was precisely her intent: Anyone she met there is more likely to be someone she wanted to meet! (Perhaps these days, "Ground Zero"—the former site of the World Trade Center—would trump all of the above.)

There are other kinds of coordination games, many of which involve various patterns of complementarity, and most of which are intentionally choreographed, to achieve a "win-win" outcome. Take, for example, Tea for Two:

		You bring	
		hot water	tea bags
I bring	hot water	− , −	+ , +
	tea bags	+ , +	− , −

Or Fast Food Picnic: You bring either hot dogs or hamburgers, I bring either ketchup or mustard. In such cases, it doesn't matter who brings which, but whatever you bring, I should bring its complement.

By the way, it should be apparent by this point that even though game theory is dead serious, there is nothing wrong—or even, especially incongruous—about treating it with a sense of playfulness. At one point, Klara von Neumann, wife of the distinguished mathematician, physicist, and game theory founding father, John von Neumann, announced with exasperation that she was tired of hearing about her husband's pet theory, and wouldn't listen to another word unless it "included an elephant." As a result, on page 64 of *Theory of Games and Economic Behavior,* an imposing tome that is the founding document of game theory, coauthored by Klara's husband, John, the image of an elephant, in fact, turns up, hidden in an otherwise obscure and arcane diagram!

There are many situations in which coordination pays, not the least being between husband and wife. But perhaps some of the most important—and least appreciated—cases involve business. So we find complementarity between computer hardware and software, between television sets and VCRs, Internet retailers and overnight delivery services, venders of spaghetti sauce and dry cleaner shops.

Complementarity can also involve more than two players. In a cartoon appearing some time ago in the *New Yorker,* two large trucks were pulled over to the side of a road, one labeled "gin," and the other, "tonic." The drivers were side by side, flagging down another, smaller vehicle, labeled "olives."

Pascal's Wager

Moving now to a higher level of complexity—and ultimately, competition—let's look at a game theory matrix for a famous philosophical/theological creation, known as Pascal's Wager. Blaise Pascal was a seventeenth-century French philosopher and mathematician who offered the following argument for why everyone should believe in God. There are two possibilities: God either exists or does not. And each person, in turn, has two options: Either believe in God or don't. From Pascal's perspective—that of a devout Catholic—there are serious and predictable consequences of each intersecting outcome; that is, distinct payoffs in each case.

Before suggesting a payoff matrix for Pascal's Wager, we need to identify a complication over the Interrupted Telephone Call and Gift of the Magi Games. In the case of Pascal's Wager, payoffs aren't simply positive, negative, or neutral, but more diverse. To keep things as simple as possible, we'll assume only four different levels of payoff, ranging from 4 (the highest, or most desired) to 1 (the lowest, or least desired). Throughout this book, this is as numerical as we'll get, and we'll use the same notation in all cases.

Here, then, is a payoff matrix for Pascal's Wager:

		God	
		exists	doesn't exist
You	believe in God	4 , 4	2 , /
	don't believe in God	1 , 1	3 , /

God could exist or not, and you could believe in Him, or not. According to Pascal, if God exists and you believe in Him, you get the highest payoff (4), eternal bliss in heaven. If God doesn't exist and you don't believe, you get the next highest (3). This is because although you don't go to heaven, you at least don't go to hell, *and* you get to enjoy the benefits of a sybaritic life in your earthly existence (no time wasted praying, going to church, dealing with rosaries or confession, and so forth). On the other hand, if you are a believer and God doesn't exist, your payoff is lower still (2), because you presumably deprive yourself of certain earthly pleasures, waste a lot of time and energy in useless religious devotion, and don't receive any heavenly reward. Finally—and most ominously—if you don't believe and God does exist, you are in deep trouble, since you suffer eternal damnation: the lowest payoff (1).

The matrix shown here also includes payoff values for God; not surprisingly, they are somewhat problematic. They also aren't important, since in this case, the decision must be made by you, the potential believer/unbeliever. Although a consistent theology might grant God the option of deciding whether or not to exist, for our purposes, this isn't God's choice: either He does, or He doesn't. The choice belongs to the human being, who must behave, however, as though God has "decided" whether to exist or not, and for whom the consequences are quite profound.*

In any event, we can still hypothesize about God's payoff. If He doesn't exist, then presumably it doesn't matter to Him whether anyone believes or not; accordingly, there is no payoff to the nonexistent column player in the upper and lower right. It can be argued that if God does exist, He is so big and you are so small that your belief or disbelief doesn't really matter to Him. On the other hand, both the Old and New Testaments make clear that the deity of these documents is jealous of people's faith in other gods, and deeply concerned about the fate of human souls; moreover, insofar as disbelief is punished by eternal damnation and belief is rewarded by heavenly bliss, it seems reasonable to conclude that one's belief or nonbelief is consequential for God as well. So we can make the following attributions: If God exists, and you don't believe in Him, God gets a very low payoff (1), presumably not as devastating to Him as your eternal damnation would be to you, but a likely cause of deep, divine disappointment, nonetheless. And similarly, if God exists and you believe in Him, let's say that God gets a very high payoff (4), again presumably not as glorious for Him as heavenly bliss would be for you, but one that reflects something between celestial satisfaction and divine delight.

These payoff values are admittedly arbitrary. Later, we'll encounter a few situations in which payoffs are precise, with 23, for example, meaning exactly 23—dollars, days of vacation, units of biological success, or some other measure of benefit. But for Pascal's Wager, and for most cases, we'll simply list payoffs 1 through 4 in order of preference, with—once again—4 being the most preferred and 1 being the least. This does not mean that a payoff of 4 (heavenly bliss) is four times

---

*As presented here, Pascal's Wager is really a decision, or a one-person "game" against nature. It is also possible to view it as a regular 2 × 2 game, in which God decides whether to reveal himself or not, as in Steven Brams's book *Theory of Moves* (Cambridge University Press, 1994).

better than a payoff of 1 (eternal damnation), just that 4 is preferred over 3 which is preferred over 2, which is preferred over 1. That is, 4 > 3 > 2 > 1.

Pascal had no doubt about the best course of action. If the payoff represented by 4 is very large and that of 1 is very small (or, hugely negative), better believe! God, concluded the great mathematician, is a good bet.

Simple Strategies

The 2 × 2 games we've examined so far don't leave much room for interaction, or strategy. Our next ones do.

In the various coordination games, both players stand to gain the same payoff in each quadrant: both win, or both lose. Their payoffs are positively linked. There are also coordination games in which the payoffs, although linked, are opposed. Take, for example, a game about the game of baseball: Pitcher-Batter. Right-hitting batters do best against left-throwing pitchers, and left-hitting batters do best against right-throwing pitchers; right-right and left-left generates a higher payoff to pitchers, whereas right-left and left-right provides a higher payoff to batters. (Why? It's a matter of physics, not game theory: A curve ball breaks to the left when thrown by a righty, and to the right when thrown by a lefty. This shouldn't matter, except that a curve ball moving toward the batter is easier to hit than one moving away.) Given this asymmetry, it isn't surprising that the team at bat is inclined to bring in a pinch hitter who will enjoy the opposite-side payoff . . . after which the pitching team can be expected to bring in a pitcher to reestablish the same-side advantage, and so forth. Canadian political scientist Thomas Flanagan[2] used batting averages as the payoffs, and constructed the following matrix, derived from the 1995 Major League baseball season:

		Pitcher is	
		left-handed	right-handed
Batter is	left-handed	.275	.276
	right-handed	.275	.264

You can read this as a symmetrical—"zero-sum"—game, since a positive payoff to the batter is precisely reflected as a negative payoff to the pitcher; in other words, the higher the batter's batting average, the lower is the pitcher's "pitching average." As expected, right-handed batters do best against left-handed pitchers, just as left-handed batters do best against right-handed pitchers; pitchers do best when it is lefty versus lefty or righty versus righty.

The strategic, game-playing component of this match-up explains why every bullpen contains both right- and left-handed pitchers, and every batting lineup consists of a mixture of right- and left-handed batters. Otherwise, either side would stack the deck in its own favor.

There is another, very similar game, often played by children and known as Matching Pennies. Here, each player tosses a penny, which lands either heads or tails. By earlier agreement, you win if, say, the two coins match; the other player wins if they are different (or *vice versa*).

		The other player tosses	
		heads	tails
You toss	heads	+ , −	− , +
	tails	− , +	+ , −

The point is that as with the Pitcher-Batter Game, one side wins if both make the same "move," and the other wins if the two are different. (Instead of righty-lefty, and so forth, it's a matter of heads-tails, etc.) Unlike Pitcher-Batter, however, Matching Pennies is purely a game of chance . . . which is why it is effective at settling disputes fairly. Each side is equally likely to win, assuming that the coins are unbiased. In Pitcher-Batter, the moves are necessarily made in sequence, with each team tempted to respond to the other's previous move. In Matching Pennies, it doesn't matter if the coins are tossed together or separately, because even though the payoff to each player depends on both moves, each is powerless, unable to influence either the opponent's action or its own.

Ratcheting things up a bit—but just a tiny bit—we come to the Goalie Game. In a number of sporting events, including soccer, hockey, or lacrosse, there are situations in which one player approaches the other team's goal with the intent of kicking, hitting, or somehow tossing a ball, puck, or whatever past the defending goalie. In the simplest case, the offensive player can go either right or left, as can the goalie. If the offensive player shoots right (or left) and the goalie moves in that direction, the shot is blocked, and the goalie wins; if the offensive player shoots right and the goalie moves left (or vice versa), the offense wins. Goalie Games are just like Matching Pennies except they involve more skill, not least because the two moves are not precisely simultaneous. The players are likely to try to mislead each other, perhaps faking one way, hoping to get the other to commit to the wrong side, and so on.

Next, a more complex matrix, in which the payoffs can be rank-ordered 1–4, as with Pascal's Wager, instead of the simple win/lose, +/– options of Pitcher-Batter or Matching Pennies. It describes a hypothetical predicament called Battle of the Sexes, first developed in a very influential book titled *Games and Decisions,* published in 1957.[3] As befits its title as well as a formulation now a half century old, Battle of the Sexes is notoriously sexist in its assumptions. But we'll retain it, not only for historical accuracy, but because its stereotypic sex role assumptions make it easier to remember.

Imagine a husband and wife out for a night on the town. The wife wants to attend the ballet. The husband prefers a boxing match. If each were alone, the decision would be easy: Each would act on his or her preference. But since they are happily married and genuinely enjoy each other's company, both prefer to attend the same entertainment, together, rather than their preferred choice, alone. Now, add a bit of theatrics, necessary to make each player's move independent of the other: The husband and wife have become separated during the afternoon. Each knows the other's preference—and also, of course, his or her own—but neither knows what the other is going to do. Each also knows that the other is likely to make a decision taking the other's decision into account! Here, then, is the Battle of the Sexes Game, using the same shorthand as for Pascal's Wager (low numbers such as 1 indicate the lowest payoff, while 4 is the highest):

		Wife goes to	
		ballet	boxing match
Husband goes to	boxing match	2 , 2	4 , 3
	ballet	3 , 4	1 , 1

The husband does best (4) at the boxing match, but only if the wife is there, too, herself receiving a somewhat lower payoff (3). By the same token, the wife does best (4) at the ballet, but only if the husband is there, too, receiving a mirror image payoff of 3. In this game, each player does worst if both of them make the sacrifice of attending the other's preferred event: the husband goes to the ballet, the wife goes to the boxing match, and each winds up doing something he and she doesn't enjoy, and, to make matters worse, alone! The next worst outcome is if both act selfishly and attend his or her preferred event; that way, even though the husband gets to watch the boxing and the wife gets her ballet, they are still alone. In summary, each does best sticking to his or her preference, *if* the other cooperates. In this case, the one who remains stubbornly insistent and selfish gets to attend the entertainment of his or her choice, and also gets the other's company.

Let's imagine the husband's thinking, based on game theory alone. (Refer to the matrix, above, if in doubt.) "My wife might go either to the ballet or the boxing match. If she chooses ballet, I'd better choose the ballet, too. After all, 3 is better than 2. But if she chooses the boxing match, I'd better go there; 4 is better than 1." The wife is in a similar quandary. Neither can decide what to do, based on the information available so far. Both husband and wife risk a kind of gridlock, with each attending his and her preferred event separately. Both would be better off if either would cooperate by attending the other's preferred activity, but the one who gives in gets a slightly *lower* payoff than the one who stands firm. This has led game theorist Anatol Rapoport to refer to such a player as a "hero." At the same time, there is this irony: If both elect to be heroic, independently, they both get the worst possible outcome![4]

On the other hand, in a real-life husband and wife situation, there may be substantial extra payoffs to being heroic, since the players are also partners and interaction between them presumably does not end with attendance at the ballet or the boxing match. One—or both—may, in fact, be eager to act the hero, in return for other benefits . . . or, at minimum, to avoid recriminations. Although this may seem to upset the game theoretic approach, it needn't do so. As is so often the case, rather than rendering the payoff matrix incorrect, this simply changes it.

To change it yet more, let's make the more realistic assumption that husband and wife still make their own choices, but in a shared context. Instead of being accidentally separated, they are together, deciding what to do. Look at the matrix again: You should be able to convince yourself that if you are either husband or wife, and you have arrived at either the upper right—both will attend the boxing match—or the lower left—both will attend the ballet—the situation would be, as game theoreticians call it, "stable." At this point, it doesn't pay either husband or wife to deviate, so long as the other doesn't. (If husband is definitely going to the boxing match, wife would rather go with him than go to the ballet alone. Similarly, if wife is definitely going to the ballet, husband would rather go, too, than attend the boxing match alone.)

The question is getting there. Each may be inclined to issue a kind of ultimatum: Husband might buy two tickets to the fight and then announce that he has done so, or wife might purchase two ballet tickets. Whoever does so first wins, since the other is likely to accede! The idea is to make an irrevocable commitment, then present the "fait accompli" to the other.

R. Duncan Luce and Howard Raiffa, who first described this particular Battle of the Sexes, conclude as follows:

> Thus, we see that it is advantageous in such a situation to disclose one's strategy first and to have a reputation for inflexibility. It is the familiar power strategy: "This is what I'm going to do; make up your mind and do what you want." If the second person acts in his own best interests, it works to the first person's advantage.

At the same time, what is logical and game theoretically "correct" isn't always psychologically astute. There is a level of defiance here that may be okay for strangers (or maybe not), but certainly isn't recommended for intimate partners. In this case, what is mathematically impeccable and logically consistent is also psychologically and romantically stupid. One moral from all this: Don't marry a game theoretician! (And perhaps, don't be too eager to *be* one, either, at least in some aspects of your personal life.)

On the other hand, there are many interactions of the Battle of the Sexes sort that do not involve husband and wife, in which reaching an agreement—any agreement—is preferable to being deadlocked. In the simplest such cases, each side wants to get its way, but both acknowledge that the worst outcome is if no agreement is reached. For example, let's imagine that a community has received a federal grant to be used for "environmental improvement" with one stipulation: If the money is not used by a given date, it reverts back to the Environmental Protection Agency. One local contingent wants to use the money to purchase additional parkland; another is sold on pollution control. Park-preferers would nonetheless rather have a sewage treatment plant than nothing at all; ditto for antipollution-proponents when it comes to increased parkland. The matter comes up for a vote in the local town meeting, whose game might look like this:

		Park-preferers vote for	
		more parkland	sewage treatment
Antipollution-proponents vote for	sewage treatment	1, 1	4 , 3
	more parkland	3 , 4	1 , 1

Here, each group is happiest (it gets 4) if the other group supports its preference, but neither side does poorly if it loses . . . so long as some agreement is reached, the worst either gets is a 3. If they vote at cross purposes, however, everyone gets a paltry 1, since the community gets neither parkland nor a sewage treatment plant. Although cooperation would certainly be best—and may well be achievable—there is nonetheless a temptation for each side to be stubborn, threatening the other side with the possibility of getting a mere 1 instead of an acceptable 3, if no agreement is reached.

In this case, as with the Battle of the Sexes, the side that gives in can be called a hero, although hard-liners on that same side might choose a more critical epithet, such as "spineless."

There is no way to predict which side will win, or, indeed, whether both might lose, if the hard-liners on each side hold sway. This uncertainty is partly a result of the fact that the games are symmetrical: In our hypothetical example, park-preferers and antipollution-proponents are equally committed to their favored outcomes, and equally (dis)satisfied if they lose. But what if the payoffs are asymmetric?

When this is true, the likelihood is that the side that cares more will win. For a familiar example, and a cultural icon at that, consider the Old Testament story of Solomon's Judgment: Two women come to King Solomon, each claiming to be the mother of the same infant. Solomon, in his wisdom, proposes that the child be cut in two, whereupon one of the women agrees not to press her suit, thereby offering to give the child to the other claimant . . . and indicating to Solomon that she is, in fact, the actual mother. Here, the payoffs to the real and pretend mother are not symmetric; even if the positive payoff of being awarded the child (alive) is equally great for both, the negative payoff of having it killed is far more severe for the real mother. You might want to construct your own payoff matrix for Solomon's Judgment, with the players being Real Mother and Fake Mother, and the "moves" being "allow division of child" and "refuse."

Of course, in this case, the real mother ends up winning, because Solomon correctly interpreted the asymmetry and its impact on the two women. (He evidently didn't count on the possibility that the fake mother might have anticipated his strategy, and also refused the division.)

In any event, it is sometimes useful to *appear* inflexible, then demonstrate flexibility at last. Consider politicians deciding to support or oppose pieces of legislation, or two corporations maneuvering to enter a new market, or leaders of the Taliban declaring—until their surrender—that they will fight to the death. In such cases, laying out one's options in this way should help clarify the possibilities, as well as the limitations of logic, while also demystifying the matrix itself.

Follow the Leader?

Considering only simple, symmetric 2 × 2 games, theorists agree that there are four different kinds that are strategically interesting. Of these, two—Prisoner's Dilemma and the Game of Chicken—have received most attention from theorists; they will from us, too (in chapters 3, 4, and 5). We've already considered one of the others, Battle of the Sexes. Let's also look briefly at the remaining one, because it's pretty nifty as well.

Called Leader, this game is most familiar when two drivers are waiting to enter heavy traffic at an intersection. A gap appears in oncoming traffic, and both would like to go, but somebody has to go first. Optimally, one should concede right of way to the other, although presumably there is a very slight benefit in being the one who goes first. But they shouldn't both concede, or neither of them goes, and the gap quickly disappears (a whiff of a coordination game). On the other hand, if both drivers try to go, the result is the worst outcome of all: a collision! Far better than that would be to concede, so long as the other goes, in which case the conceder gets onto the highway, just a bit after the leader. Summarizing: Follow the leader is a good strategy; being the leader is even better; waiting, fruitlessly, for each to be the leader is a poor strategy; both trying to be the leader is the worst of all.

The Game of Leader might look something like this, once again with higher payoff numbers being good and lower ones being bad:

		Driver 2	
		goes	concedes right-of-way
Driver 1	goes	1, 1	4 , 3
	concedes right-of-way	3 , 4	2, 2

What is the best move for driver 1? As with all games, it depends on what the other player does. If driver 2 decides to go, then driver 1 should concede the right-of-way (3 is a better payoff than 1). On the other hand, if driver 2 concedes the right-of-way, then driver 1 should go (4 is a better payoff than 2).

Now, let's make a change in terms. For *go*, substitute *defect* and for *concede right-of-way*, substitute *cooperate*. This is consistent with the logic of these actions, and with much of game theory, since going straight ahead, seeking one's own benefit regardless of the possible cost to the other player, represents a kind of defection from norms of social nicety. And likewise, to concede the right-of-way is to behave cooperatively:

		Driver 2	
		defects	cooperates
Driver 1	defects	1 , 1	4 , 3
	cooperates	3 , 4	2, 2

Each player is best off being leader (defecting) so long as the other decides to be a follower (to cooperate). But two leaders (two defectors) equals two losers. So does two followers (cooperators) . . . although they lose considerably less. If both players play it safe—and make sure that they avoid the worst possible outcome—each cooperates, preferring to be a follower. As a result, there will be no collision, but also no movement! This, however, is not a stable equilibrium, because each would have reason to change from his chosen posture of immobility as soon as he or she knew that the other was waiting.

Actually, there are two equilibrium points here: the upper-right corner and the lower left. Once it is known, for example, that driver 1 is going ahead, the best move for driver 2 is to follow. Similarly, if it is known that driver 2 is going ahead, the best move for driver 1 is to follow. (Recall the Battle of the Sexes.) The problem—as with Battle of the Sexes—is that the system isn't stable, since each would slightly prefer to go ahead and have the other follow. This is one reason why rules for automotive right-of-way are a good idea.

Battle of the Sexes and Leader are similar but not identical. In Battle of the Sexes, the worst payoff arises if both players cooperate, so that the husband ends up alone at the ballet and the wife ends up alone at the boxing match; the next worst is for each to defect, with the husband attending the boxing match, alone once again, while the wife attends the ballet, also by herself. In Leader, the worst outcome is for both to defect and go straight (into each other); the next worst arises if both cooperate, and as a result, no one moves.

In real-life "games" of Leader, right-of-way is determined in many different ways, some of them legally enshrined, some not: who arrives at the intersection first; which street is more prominent; aggressive or conciliatory hand or facial gestures; flashing of headlights, honking of horns, barely discernible inching forward; whether one person is driving a small compact car and the other an imposing SUV. Indeed, part of the antagonism toward the latter may well derive not just from their gas-guzzling, but also from the degree to which they embody the threat of on-the-road defection.

Guessing Games

There was a time when people looked to game theory as to an oracle or Fount of Infinite Wisdom: "What shall I do, O Master, so as to confound my enemies, or, at least, to get the best payoff for myself?" In a very limited number of cases, game theory has come through. Interestingly—and fitting the penchant for paradox so characteristic of the enterprise—in those relatively few situations for which game theory does, in fact, provide specific answers, the recommendations tend to be couched in terms of probabilities rather than specifics. Rather than "Do this" or "Do that," the Master Matrix is more likely to advise: "Do this, on average, with probability X—in other words, X percent of the time." (And therefore, "Do that, on average, with probability Y— in other words Y percent of the time," with X plus Y equal to 1, or 100 percent.)

Take, for example, the Coin Matching Game we have already encountered.

		The other player tosses	
		heads	tails
You toss	heads	+ , −	− , +
	tails	− , +	+ , −

You win if your toss matches the other player's; the other player wins if the two are different. There is no simple advice for how to win such a game. Even if you could determine whether your coin would land heads or tails, there is no guaranteed benefit either way, since the payoff depends equally on what the other player does, and you have no control over that. Let's imagine, however, that you could control your own moves, since that makes the game much more interesting. Instead of flipping a coin, you *choose* whether to "play" heads or tails; then you present your move to the other player, who has gotten to make the same choice, independently. What, now, is your best move?

Once again, there is no answer. If you are playing the Controllable Coin Matching Game only once, it doesn't matter which you choose. Heads and tails are equally likely to make you victorious. But what if you are playing the game many times? In this case, there is, in fact, a best strategy: You should play heads and tails equally often—that is, with an average probability of 50 percent. It is a "winning strategy" only in the most pessimistic and limited manner: it isn't a guaranteed winner, but it is guaranteed that any departure from this 50:50 strategy will be a loser!

If you get in the habit of playing heads more than one-half the time, the other player will soon catch on, and start playing tails more than half the time. Conversely, if you bias your play toward tails, the other player will overbalance toward heads. Either way, on average, you lose.

Even if you had a completely controllable coin, which would obediently land heads or tails as you wished, the best way to direct it would be to flip a coin! (A different one—one that is "honest," which is to say, equally likely to land heads as tails.) It seems to violate the laws of logic that the optimum strategy in such a case is to rely on chance, but there it is. However, this isn't really a matter of casting all rationality to the winds and making one's play an entirely random "nondecision." Instead, the limits of probability are mindfully decided—in this example, 50:50—and then, the decision rule is allowed to work its will for any individual move.

Nonetheless, it seems strange that in some cases, the best strategy is to rely on chance. Stranger yet when these cases are those that seem to call for the most informed judgments. Military or business leaders, for example, would not enhance their reputation if it became known that they literally flipped a coin or rolled some dice when it came to making certain crucial decisions. After all, that is what they are paid to do: to make *decisions*, presumably based on their superior grasp of the situation, not on blind chance. The point is, however, that sometimes a superior grasp of the situation suggests that blind chance—within certain parameters—is the best strategy.

It is also worth pointing out that the best strategy for games of this sort isn't *always* a probabilistic 50:50. If the payoffs are unbalanced, the "winning" strategy will be comparably unbalanced.

Professor Anatol Rapoport[5] gives this example. Call it the Circular Thwarting Game. [Warning: It is worth following, not only for the pleasure of seeing the alternatives unfold, but also as an example of game theory's labyrinthine logic. However, if you don't enjoy this sort of thing, or find it just plain confusing, feel free to skip ahead; this particular game isn't essential for understanding the basic material of the book.]

		Individual B chooses	
		b1	b2
Individual A chooses	a1	0, 0	+ 2, −2
	a2	+3, −3	−1, +1

As with Coin Matching, this is known as a zero-sum game, since whatever individual A wins, B loses, and *vice versa*. Each player has two choices: A can choose a1 or a2, and B can choose b1 or b2. Individual A would like to win +3, so he is tempted to play a2. But he'll only get that if individual B plays b1, which is not likely to happen! In fact, individual B, looking at the same matrix and anticipating that A might try for this payoff, could play b2, whereupon individual A would get only −1 and B would get a positive payoff of +1. For individual A, therefore, knowing that B will likely try to thwart him, a1 is a safer bet. That way, he might get 2 (if B plays b2) but can't get worse than 0.

How would B look at it? He would like to win +1, and can do so by playing b2, but he can't count on individual A playing a2. In fact, given individual A's reasoning (just described), individual B suspects that A will play a1. In that case, B should play b1, assuring that he won't get stuck with −2. But things can get more complex yet. B may well figure that A will figure that B will be choosing b1, in which case A might choose a2 after all, hoping to win +3! If so, then B could outfox A by choosing b2 anyhow, and getting +1 for himself and sticking A with a measly −1. But if A has been anticipating B's moves to this point, A might well choose a1 after all, so that B should pick b1. There is no end to it.

There is an answer, however, one that applies if the game is repeated (or, as game theoreticians put it, "iterated") rather than simply played once. In this case, the best strategy would be for A to play a1 two-thirds of the time and a2 one-third of the time, and for B to play b1 and b2 one-half the time. This way, each one is assured a payoff that is the least bad: A always gets +1 and B always gets −1. Consider what happens: When B plays b1, A gets 0 two-thirds of the time (because he plays a1 two-thirds of the time) and A also gets 3 one-third of the time, for a total of +1. When B plays b2, A wins 2 two-thirds of the time, for a total of four-thirds, and he also loses 1 one-third of the time, for a loss of one-third. Add these together, and A ends up, again, with +1 on average. Similarly, B assures himself of −1 (not a great payoff, but given the unfair matrix, which is already biased toward A, not too shabby).

The best strategy in such a case, as with the much simpler Controllable Coin Matching Game, is to mix your moves according to a particular probability; in this case, one-third and two-thirds (one-half and one-half for the Coin Matching Game). If payoffs differ or if there are more than two possible moves, the preferred probabilities will also shift. Here is a familiar example: the Paper-Rock-Scissors Game often used by children to settle disputes. It is more complex than most of the games we'll be considering from here on, since each player has three possible moves, but otherwise, it is very simple indeed. Each player makes one of three moves: displaying one's hand so as to indicate Paper (flat hand), Rock (fist), or Scissors (first and second fingers extended). The decision rule is circular: Paper defeats (covers) Rock, Rock defeats (breaks) Scissors, and Scissors defeats (cuts) Paper. Here is a payoff matrix for the Paper-Rock-Scissors Game:

		The other player chooses		
		paper	rock	scissors
	paper	0, 0	+, −	−, +
You choose	rock	−, +	0, 0	+, −
	scissors	+, −	−, +	0, 0

As with the coin game, there is no one best move when playing Paper-Rock-Scissors. There is, however, a best *strategy*: Choose each of the three moves with equal probability, 33.3 percent of the time. Once again, winning is not guaranteed, but you can be certain there is no better way to play the game.

Next, let's go back to the Controllable Coin Matching game (because its 2 × 2 matrix is a whole lot friendlier than a 3 × 3). What happens if we vary the payoffs and make them asymmetric? In fact, very asymmetric? Imagine that if you and the other player both play heads, you win *$1 million* (and he loses $1 million), whereas for all other combinations, just one penny changes hands, using the same pattern as before. The new matrix would be like this:

		The other player chooses	
		heads	tails
	heads	+$1 million, −$1 million	−1c, +1c
You choose	tails	−1c, +1c	+1c, −1c

It seems obvious: The best strategy is for you to play heads, over and over. After all, you stand to gain $1 million if only the other player would also play heads, just once! But the other fellow understands that playing heads is playing into your hands. So he'll keep doing tails. And whenever you play heads and he plays tails, you lose a penny to him. Knowing that, the best move is for you to play tails, too. So you give up a cool $1 million and settle—rationally and logically and game-theoretically, albeit probably a bit regretfully—for a penny instead. So long as everyone knows the matrix, this game is loaded in your favor, but not by very much! Just a penny.

This result is an interesting paradox: By increasing the payoff for a particular move—so that in our case you can possibly get $1 million— the result is actually to decrease your likelihood of choosing it! Initially, you may well select heads, over and over. But your opponent, if rational, will always choose tails, so that after a while, you come to realize that choosing heads is, in fact, irrational, and that the best you can realistically hope for is to choose tails and get a piddling penny.

But the story isn't quite over. If this game goes on long enough, the other player may well get tired of paying you a penny every time. Since he can be confident that you will play tails (because you are ruefully aware that he will also play tails), it might pay him, *very rarely,* to play heads, in which case you'll have to pay him a penny! But he won't want to do that very often, since if you get even a hint that he might do this, it would pay you to try heads every once in a while, in hope of getting the $1 million. It turns out that if you played this game over and over, it would pay you to try heads very, very rarely, about one play in 50 million. (This can be calculated by a simple game theory formula . . . which is, nonetheless, too complex and not really necessary for our purposes.)

What we've been looking at is a situation in which once again, the best way to play the game is to vary your moves according to certain probabilities, in this case intended to "keep the other side honest."

It happens in real life, too. In football, even if a team has a very strong running game, which can be counted on, it will occasionally throw a forward pass, to keep the other side honest, so they don't expect a running play every time and load their defense in expectation of it. Similarly, in baseball, most pitchers mix up curves, fastballs, sliders, and so forth. This is because the batter has a real advantage if he knows what is coming. (It was revealed in early 2001 that during the renowned 1951 playoff game between the New York Giants and the Brooklyn Dodgers, won by a last-minute, ninth-inning home run on the part of the Giants' Bobby Thomson, a spy in the bleachers—who intercepted signals sent by the catcher—gave Thomson advance warning that the Dodgers' Ralph Branca was going to throw a particular type of pitch.)

Similarly with poker. Only a very poor poker player would let his betting be determined entirely by the quality of his cards. Even with perfect knowledge of the probabilities of drawing, say, three of a kind, given your current hand, it would be disastrous to bet accordingly. And worse yet to bet high only when you actually have a strong hand. This is because you would automatically telegraph your holding: if you follow simple logic and bet high when you have strong cards, and only then, others would drop out every time, so you would win very little. By the same token, whenever you bet low, others would know you have a weak hand. So, the optimum strategy is to bluff at least sometimes; this way, when you have a strong hand, the others won't know that you do. Some may figure you are bluffing and stay in. Again, what we see here is the value of keeping the other player(s) honest, by making them uncertain as to what you are going to do next.

There are some interesting exceptions, however: If you have an absolutely devastating fastball, then you might be able to get away with throwing it time after time. More interesting for our purposes: If you are a knuckleball artist, you can throw all knucklers . . . because a knuckleball, by definition, is unpredictable!

An effective way to keep animals inside a fence line is to electrify it, with a "hot wire" that provides a mild but distasteful shock. It isn't necessary for the wire to be "hot" all the time: One or two aversive experiences are usually enough to induce most creatures to respect it, although here, too, the rancher can't get away without backing up his threat, by at least occasionally—and unpredictably—turning on the power. Or think about the warning sign that announces "This property patrolled by a vicious guard dog, three days a week. You guess which days!"

The Battle of the Bismarck Sea

Game theory is especially attractive to military strategists. Why? Because war is often a kind of game, not in the lighthearted sense, but because war itself—as well as the battles that make up most wars—generally involves just two opponents, both of whom are genuinely opposed. Often, in fact, the situation is a zero-sum game, since if one side wins, the other loses. (In other words, the sum of everyone's payoff is zero.) In addition, a successful general or admiral needs to take the other side's likely course of action *very* seriously, and to plan accordingly. Only rarely does a commander announce, "Damn the torpedoes, full speed ahead." The best strategists, in fact, are notable for the care with which they take the other side's behavior into account: "If I do this, my opposite number is likely to do that, in which case I should then do this . . ."

Here is President John F. Kennedy's adviser, Theodore Sorenson, describing some of the mental machinations that took place in Washington during the Cuban Missile Crisis of 1962: "We discussed what the Soviet reaction would be to any possible move by the United States, what our reaction with them would have to be to that Soviet reaction, and so on, trying to follow each of those roads to their ultimate conclusion."[6]

Presumably, similar "gaming" was going on within the Kremlin.

Once again, the gamey quality in all this is revealed by the fact that the outcome for each side depends not only on what it does, but also on how the other acts, as well as the degree to which each side, despite (or, maybe, because of) being unable to control the other's actions, attempts to anticipate what the other is likely to do. Sometimes, the results can get a little silly, as each tries to assess what the other is planning to do, which in turn influences what it will do, which the other also supposedly takes into account, and so forth. The following description appeared in the *London Daily Express,* shortly after the Battle of the Bulge, in 1945:

> The Allies were not surprised, because they knew the possibility of a surprise attack. What surprised them was that the Germans thought it worthwhile to make a surprise attack in spite of the fact that such an attack, though deemed possible, was not deemed probable, in view of the fact that we knew they would try to surprise us.[7]

In other words, the Allies thought that the Germans might attempt a surprise attack, but knowing that the Germans knew that the Allies expected it, the Allies assumed that the Germans probably wouldn't do it! But they did.

Game theory thinking isn't only for theoreticians or Monday-morning, after-the-fact quarterbacks. It is also the stuff of real-life decision making during war. One of the best analyzed of such cases occurred during the Battle of the Bismarck Sea, during World War II, and was described in a famous paper by military historian and game theoretician O. G. Haywood Jr.[8] It is a genuine two-person zero-sum game.

Here is the scene. There are two players: the American general George C. Kenney, commander of Allied forces in the South Pacific, and Japanese admiral Hitoshi Imamura. It was February 1943, at a critical time during the struggle for New Guinea. Allied intelligence learned that a large Japanese convoy was assembling at Rabaul, a port in the nearby island of New Britain; its goal was to reinforce the Japanese position in New Guinea. Kenney's job was to attack the Japanese convoy; Imamura's was to direct it safely to its destination. Each "player" had to choose one of two "moves." Imamura could direct his convoy to sail either north of the Pacific island of New Britain, or south. Either route would take three days, but rain and poor visibility were very likely on the northern route whereas good weather was likely on the southern. Imamura's main concern was to minimize the likelihood that Kenney's forces would encounter his.

Kenney, for his part, had to decide whether to concentrate his reconnaissance aircraft on the northern or southern route, bearing in mind his goal of maximizing the likelihood that his forces would detect Imamura's. Finally, add this crucial bit of information: Kenney's staff estimated that if his reconnaissance aircraft flew north, it would take a day to sight the Japanese convoy, regardless of whether Imamura chose north or south. The result would then be two days of bombing. But if the Allied airplanes flew south, then Kenney's forces would have the opportunity of either three days of bombing (if Imamura also chose the south), or just one day (if Imamura chose the north).

The payoffs are measured in numbers of days of bombing available to the Allies, who want this number to be as high as possible, while the Japanese want it to be as low as possible. One day of bombing is comparatively good for the Japanese, bad for the Allies; three days of bombing is good for the Allies, bad for the Japanese. Two days is medium for both. The "game" is zero-sum since a positive payoff for one is reciprocally negative for the other.

Here is a payoff matrix for the Battle of the Bismarck Sea:

		Imamura (Japanese)	
		sends convoy north	sends convoy south
Kenney (Allies)	searches north	2, −2	2, −2
	searches south	1, −1	3, −3

Imamura and Kenney are each faced with the classic game theory problem of making their decisions without knowing what the other is doing, but also knowing that outcomes depend on the other's behavior no less than on one's own, as well as the fact that each is trying to out-fox the other. In addition, the Battle of the Bismarck Sea is unusual in that it is a simple zero-sum situation; days of bombing that are a positive return to Kenney are exactly negative to Imamura (+1 for Kenney, for example, equates to −1 for Imamura, and so forth).

Let's look first at Kenney's options. If he searches north, he is guaranteed two days of bombing (+2) regardless of whether Imamura has gone north or south. Searching to the south, on the other hand, is riskier for Kenney: He gets a better payoff (+3, or three days of bombing) if Imamura has elected to go south, but a worse payoff (+1, or only one day of bombing), if Imamura has gone north.

There is no easy-to-determine best strategy for Kenney. It depends on whether he is willing to take a chance—hope for three days of bombing and run the risk of getting only one—in which case he should search south, or whether he decides to guarantee a minimum gain, in which case he should search north. By choosing north, Kenney is assured that he won't get less than two days payoff, regardless of what Imamura has done. Game theorists call this his "maximin" strategy, because it guarantees him the maximum of his possible minimum outcomes. It is a way of cutting his losses, making sure in this case that his forces won't get fewer than two days of bombing time. (Because the situation is biased in his favor, Kenney is actually guaranteed to come out ahead in any event; he nonetheless wants to assure that the Allies get as much gain as possible.)

Now, let's look at Imamura's options. Although perfectly symmetrical to Kenney's, they are much less attractive, because this real-life game is loaded in the Allies' favor. No matter which choices are made, the Allies will get to do some bombing, once they receive correct intelligence information as to the Japanese plans. Either way, therefore, Imamura is going to lose something. His goal is to minimize that loss. In game theory terms, he is likely to play a "minimax" strategy, trying to minimize the maximum payoff to his opponent . . . or, one might say, minimize his own maximum loss. From Imamura's perspective, those losses will amount to either −2 or −1 if he goes north (−2 if Kenney searches north, and −1 if Kenney searches south), *versus* either −2 or −3 if he goes south (−2 if Kenney searches north, and −3 if Kenney searches south). Seen this way, Imamura's best (minimax) strategy is to go north, since that way, he can't do worse than a −2, and may even get a mere −1.

What actually happened? Kenney and Imamura chose their maximin and minimax strategies, respectively. Imamura went north and Kenney searched north. In the real Battle of the Bismarck Sea, the Japanese suffered their most serious defeat in the South Pacific up to that time, losing ten warships, twelve transports, and fifteen thousand men. Neither commander made a strategic error, however: if Kenney had searched south instead of north, he would have provided his forces with only one day of bombing instead of the two they enjoyed. And if Imamura had proceeded south, he still would have suffered two days of bombing. (Had he gone south and if Kenney had simultaneously searched south, Imamura would have suffered a worse outcome yet: three days of bombing, or −3 for him and +3 for Kenney.)

In the Battle of the Bismarck Sea, Kenney's and Imamura's optimal strategies intersected at a particular equilibrium, also called a "saddle point," for the following reason: Think of a payoff matrix being like a horse wearing a saddle. Looking head-on at a horse, the saddle sits at the animal's highest point, his back. This is the perspective of one player, who wants the highest payoff. Looking at the same horse from the side, the saddle is located at the animal's lowest point between its head and its haunches, just behind the crest of its neck. This is the perspective of the player who wants to minimize the amount given away to the other player. (It is also the situation when, as writer Italo Calvino once put it, "You know that the best you can expect is to avoid the worst."[9]) It is no coincidence that saddles are normally placed where they are, and not high up on a horse's neck, or directly above its hind legs. Saddle points, on a horse, are stable. Also in game theory.

Every horse has a saddle point; not so for games. But when they do, they are usually the best place to be, for both participants. Even here, though, things are a bit tricky. In the Battle of the Bismarck Sea, Kenney had no way of knowing Imamura's plans, except to assume that the Japanese admiral might well be choosing whatever strategy offered minimal risk to his own forces. But if Kenney had reason to suspect that Imamura was in fact going south—perhaps because Imamura was known to have an irrational preference for southern headings—then he (Kenney) would have been justified in going south, too.

There is something especially unpleasant about zero-sum games, if only because they bespeak a situation of "perfect" (or perfectly awful!) competition, dog eat dog and winner take all. But they do happen, and not only in war: Take, for example, an election, in which there can be only one winner, two corporations competing for a single contract, two nations competing for an alliance with a third, two silverback male gorillas competing for the sexual favors of a female, and so forth. A strong case can be made, in fact, that many biological situations are indeed zero-sum, insofar as there is a limited amount of food, nesting territories, prospective mates, and so forth, and yet, because every population has the potential of geometric increase, a nearly unlimited number of individuals compete for these finite resources.

Nonetheless, perfect zero-sum games, like the Battle of the Bismarck Sea, are probably rare, since they imply that the two players have no common interests *at all,* whereas, in fact, even in strongly competitive situations, two sides often have at least some shared concerns. Even those sexually mature, highly competitive male gorillas may be relatives, which would give each a genetic stake in the fate of the other; moreover, each has a shared interest in not pushing the other too far, since both may well be better off if they could part uninjured and, thus, retain the prospect of breeding another day.

So, *non*–zero sum games are generally of particular importance, since they involve cases in which the players have some competitive issues but also some cooperative ones. For example, even in cases of war—extreme competition—the two sides often have a shared interest in sparing items of great cultural value, in treating prisoners of war according to certain codes, in avoiding use of poison gas, in respecting cease-fire agreements, and so forth.

In the pages to come, we'll spend most of our time outside the Bismarck Sea, looking instead at non–zero sum situations, in which there is no simple saddle point, and in which the two sides have at least the prospect of mutual gain as well as mutual pain. Nonetheless, we'll encounter battles aplenty, and challenges to rationality and well-being that are not only widely applicable and important, but also devilishly paradoxical, tantalizing, and, sometimes, downright mind-boggling.

3

Prisoner's Dilemma and the Problem of Cooperation

Football coaching legend Vince Lombardi once noted that a tie is about as satisfying as kissing your sister. He also maintained that winning isn't everything . . . it's the *only* thing! For Mr. Lombardi, life is simple and the best precepts are straightforward. Try your hardest. Aim to win. Failure isn't an option. And so forth.

But it ain't necessarily so.

Like it or not, there are situations in which an all-out commitment to victory can be, paradoxically, a ticket to failure and frustration. We are all, on occasion, imprisoned by this dilemma: trying to achieve the best possible outcome for ourselves, we bump up against comparable efforts on the part of someone else who is trying to do precisely the same thing, and, as a result, both parties end up doing worse than if everyone had figured out how to cooperate and share the benefits.

It is called the Prisoner's Dilemma, and nothing in game theory is more compelling, aggravating, or relevant. Tons of ink have been spilled over it. It occupies entire library shelves and the minds of some of the world's best thinkers. It will also be the subject of this chapter, and—in slightly modified form—the next.

The Classic Formulation

Imagine two prisoners, apprehended by the police and accused of collaborating in a bank robbery. The prosecutor's goal is to get both prisoners to plead guilty. The prisoners' goal is to get off with the lightest possible sentence; better yet, to go free. The prosecutor, however, is especially fiendish. He separates the two prisoners—keeping each in a different cell, allowing no communication between them—and makes the following offer to each. "If you and your partner continue to plead not guilty, I do not have enough evidence to convict either of you of bank robbery, but you will both be found guilty of a lesser charge, say, illegal possession of a weapon, and you will both be sentenced to a year in jail. If both of you plead guilty, then you will each get ten years in jail. On the other hand, if you cooperate with me and plead guilty to the robbery charge, thereby also implicating the other guy, then this will give me enough evidence to convict your buddy, and the government will reward you for your assistance by letting you go free while he will then be sentenced to thirty years behind bars. Of course, I am making this same offer to him: If you do not plead guilty and your partner does, then you get thirty years, and he goes free."

These options are indicated in the following payoff matrix, with each individual free to choose between two courses of action: pleading guilty or not guilty. As usual, the payoffs to individual 1 are shown first, before each comma; those for individual 2 come after. Since in this case, a larger number means more time in prison and thus, a worse payoff, they are presented as negative numbers (−30 is the worst payoff, followed by −10, −1, and finally, 0, the best payoff). This is not a zero-sum game, in which one player's gain means the other's loss; it is, in fact, negative-sum, since it is strongly biased against the prisoners. Nonetheless, each player is given the option—or, is manipulated—to choose the "move" that provides the least costly outcome. Here are those outcomes.

		Prisoner 2 pleads	
		not guilty	guilty
Prisoner 1 pleads	not guilty	−1, −1	−30, 0
	guilty	0, −30	−10, −10

Under these conditions, each prisoner faces a cruel dilemma (hence, "Prisoner's Dilemma"). Both would do best—just one year in jail—if they cooperated with each other and stuck to their plea of not guilty, the payoff in the upper-left corner. But each one fears what the other might do . . . as well as being tempted by the prospect of an even higher payoff. To understand this, let's eavesdrop on the thoughts of individual 1: "My partner must plead either guilty or not guilty. If he pleads guilty, and I also plead guilty, then I get ten years; if he pleads guilty and I plead not guilty, then I get thirty years. So, clearly, if he chooses guilty, my best choice is to plead guilty, too. (Ten years in jail is a bad payoff, but it's better than thirty.) On the other hand, if my fellow prisoner pleads not guilty, then my best choice still is to plead guilty, because then I get to go free rather than spend a year in jail."

To make matters worse, prisoner 1 is also aware that prisoner 2 is likely to be thinking similarly.

As a result, prisoner 1 finds himself forced to plead guilty out of fear that the other will double-cross him and plead guilty, combined with the allure of double-crossing the other if prisoner 2 pleads not guilty. And of course, prisoner 2, reasoning identically, comes to the same conclusion. The ironic result of this process is that both prisoners plead guilty and spend ten years in jail (the lower-right corner payoff), whereas they each could have gotten a better payoff (just one year in jail, upper-left), if they had only figured out some way of cooperating.

We can also look at this particular Prisoner's Dilemma matrix using the relative values of 1–4 introduced in the previous chapter, where 1 is the worst payoff, 2 is next worst, 3 is somewhat better, and 4 is the best:

		Prisoner 2 pleads	
		not guilty	guilty
Prisoner 1 pleads	not guilty	3, 3	1, 4
	guilty	4, 1	2, 2

Once again, the best combined payoff for both (a very short jail term, payoff 3) comes from a shared not-guilty plea, but the best for each, individually, arises if he pleads guilty (and goes free, payoff 4) while the other maintains his innocence (and, not incidentally, gets a stiff jail term, payoff 1). As a result, each is driven to plead guilty, and, as a result, each gets a poor payoff: 2, which in this case is ten years in jail.

The special devilishness of Prisoner's Dilemma is that it pays to defect if you think the other player will cooperate. At the same time, it pays to defect if you think the other player will defect. So both players get to spend an additional nine years behind bars contemplating the merits of being so clever!

This, then, is a dramatic example of how attempting to win results in a substantial loss, if both players are similarly motivated. Vince Lombardi might not have liked it, but it is all perfectly rational, so long as the payoffs are as shown and both parties are making the same effort. To summarize and simplify, the logic goes like this: If the other guy defects, you'd be a fool not to do the same; if the other guy doesn't defect, you'd make out like a bandit if *you* defect. Either way, the "correct" response—for both players—is to defect.

Let's look at the 2 × 2 Prisoner's Dilemma matrix a bit differently, changing the labeling once again, but not the logic. When either prisoner pleads not guilty, he is trying to keep faith with the other, seeking to obtain a good, shared payoff; it is an attempt at cooperation. Hence, instead of "not guilty," we can substitute "cooperate." (This has the advantage of being generalized to other situations, as we shall see.) Game theorists sometimes even use the word *nice*.

By the same token, an individual who pleads "guilty" is trying to take advantage of the other (or possibly seeking to protect himself from the other doing the same thing). To plead guilty, in this kind of situation, is to "defect" from the common good, in pursuit of one's own shortsighted benefit. Game theorists sometimes use the word *nasty*.

And finally, we can also rename the payoffs, as game theorists typically do. The highest return in this type of game, received if you defect when the other player cooperates, is often labeled T, for Temptation: the Temptation to be nasty. Next highest is R, for Reward: the Reward of mutual cooperation. It isn't as high as T, but is pretty good, nonetheless. Then comes P, for Punishment: the Punishment associated with mutual defection. And lowest of all is S, the Sucker's payoff: what you get if you cooperate only to have the other player defect.

Put these all together, and we get the following matrix for a generalized Prisoner's Dilemma:

		Individual 2	
		cooperates (is nice)	defects (is nasty)
Individual 1	cooperates (is nice)	R, R	S, T
	defects (is nasty)	T, S	P, P

A 2 × 2 game becomes a Prisoner's Dilemma any time the payoffs are in this relationship: T is largest, then comes R, followed by P, and then the smallest, S. Or, in mathematical notation: $T > R > P > S$.*

To repeat: Under these conditions, each individual is sorely tempted to defect (and get T), but if both do so, they get P. The dilemma is that although P isn't the worst possible return (that dubious honor goes to S, the Sucker's payoff), it is considerably worse than both would receive if they had cooperated and each ended up with R, the Reward for such cooperation. By following impeccable logic, motivated by the Temptation of individual gain plus the fear of being a Sucker, both players end up with a crummy outcome.

The story of the Prisoner's Dilemma made its public debut during a seminar on game theory that took place in the Psychology Department at Stanford University in 1950. It was presented, as a purely theoretical proposition, by Albert W. Tucker, a mathematician visiting from Princeton. And Tucker, it appears, was reporting the concept originally developed by theoreticians Merrill Flood and Melvin Dresher at the

*There is, in fact, a further technical requirement, that $R < (T + S)/2$, which simply means that the payoff of mutual cooperation must be less than the average payoff if the two players alternated defecting and cooperating. This stipulation is only relevant in repeated games of Prisoner's Dilemma, as opposed to the one-shot variety described above; in addition, it's very rarely a realistic possibility.

RAND Corporation some time previously. But for all its academic pedigree, the Prisoner's Dilemma isn't simply a rarified theoretical construct. In fact, something very much like a Prisoner's Dilemma arises frequently in cases of criminal prosecution, and, as we shall see, it has had substantial impact on many current dilemmas, from price fixing to nuclear weapons.

In his book *Prisoner's Dilemma,* William Poundstone points out that plea bargaining may, in fact, often involve something worrisomely close to Professor Tucker's ivory tower supposition. (Worrisome because the goal of the criminal justice system is supposed to be justice, not simply obtaining a conviction. Accordingly, it is—or should be—deeply troubling if individuals are being convicted because their confederates are arm-twisted into false confessions, motivated by fear of being a Sucker combined with the Temptation of getting off easy.) To make matters worse, Prisoner's Dilemma situations are especially likely to arise in the most weighty prosecutions, those involving the death penalty, since it is in these cases that the government needs to make an unusually strong argument. Because the standards of proof are especially high in capital cases, convictions often depend on someone testifying against someone else; that is, an accomplice "turning state's evidence," and thereby defecting from his buddies, hoping to get a better deal for himself. And not surprisingly, prosecutors in capital cases are especially likely to place the defendants in just such a dilemma.

Poundstone quotes a *Los Angeles Times* article about the death sentence of convicted child murderer Robert Alton Harris:

> In robberies where murders occur, for example, there is often more than one criminal involved and thus more than one person who may be eligible for the death penalty. . . . Frequently, a race ensues in which the robbers try to be the first to point the finger at an accomplice and make a deal with the prosecutor to testify in return for leniency. Sometimes . . . it never becomes clear whether the person who got leniency or the person on trial for his life actually pulled the trigger.[1]

In Edgar Allan Poe's short story "The Mystery of Marie Roget," Poe's detective Dupin describes what happens when members of a gang are given the temptation of defecting, of being the first—and, thus, the best rewarded—member to rat on the others: "Each one of a gang, so placed, is not so much greedy of reward, or anxious for escape as fearful of betrayal. He betrays eagerly and early that he may not himself be betrayed."

Criminal defense attorneys tell me that in practice, defendants aren't quite that eager to plead guilty and implicate their colleague(s). This is less because of honor among thieves than because defendants often fear that they or their family will be subject to retribution by friends or family of the codefendant who has been left to "take the rap." (Hence, witness protection programs.) In such cases, nonetheless, prisoners may well judge that defection—that is, pleading guilty and implicating others in hope of receiving a lighter sentence—isn't such a good strategy after all. None of this, by the way, undermines the logical structure of the game; it simply means that if the risk of retaliation is high enough, the Temptation to defect is that much lower, so that in cases of serious downside risk—for example, defecting against the Mafia—the "game" may even cease to be a Prisoner's Dilemma.

An almost perfect Prisoner's Dilemma unfolded during the Watergate investigations. G. Gordon Liddy (counsel to the Committee to Reelect the President) and John Dean, counsel to the president, were the two "prisoners." Assistant Attorney Earl Silbert needed to get the cooperation of one or the other, because otherwise he couldn't convict either. Most important, without either Liddy or Dean confessing, he couldn't get at the higher players, notably President Richard Nixon. His strategy was to implicate medium-level officials such as Liddy or Dean, who would then be granted immunity in return for testifying against those at the next level. Both Liddy and Dean were, in fact, guilty, but Liddy in particular was stubbornly refusing to plead guilty (that is, he wouldn't defect). Here's the matrix:

		Liddy	
		cooperates	defects
Dean	cooperates	cover-up works	Dean convicted
	defects	Liddy convicted	cover-up falls apart

Liddy wouldn't play the game. But Dean couldn't know this for sure. The government, recognizing that Dean was the weakest link in the cover-up chain, decided to use a Prisoner's Dilemma strategy to get him to defect, so that Liddy at least could be convicted. (Bear in mind that in this case, as with the classic Prisoner's Dilemma formulation, to "cooperate" is to cooperate *with the other prisoner,* not with the authorities. And similarly, to "defect" is to rat on one's fellow defendant . . . and thus, to cooperate *with the prosecutor.*)

Silbert brilliantly orchestrated things so as to look as though Liddy was about to defect. Thus, Liddy was kept in the grand jury anteroom for a long time, making it appear that he was turning state's evidence. Silbert had previously told Dean that Liddy had been "talking to us privately." In addition, Silbert's staff asked Liddy's lawyer to tell the press that Liddy was cooperating. He refused and did as Silbert had secretly hoped: he announced loudly to reporters that his client was keeping silent. This was interpreted by Dean as "protesting too much," and so he—convinced that Liddy was, in fact, spilling the beans—hurried to do so first![2]

If Prisoner's Dilemma applied only to criminal prosecutions or similar maneuvering, it would warrant attention, since it may well represent an unfair manipulation of defendants. But as we shall see next, its importance goes far beyond the courtroom.

A Few Examples

Picture two gunfighters, each carrying a loaded weapon, facing each other. It's a dangerous situation to be in, even if neither "player" is especially bloodthirsty, and if both would rather not shoot at all. The mutually rewarding outcome—the Reward for mutual cooperation, or payoff R—is almost certainly the best that can be achieved, for both gunfighters taken together. But each is faced with a potentially lethal dilemma: if he cooperates and refrains from using his weapon, the other might be tempted to draw first. This Temptation—payoff T in the Prisoner's Dilemma matrix—is fueled by the prospect of getting the drop on one's opponent, as well as fear that, otherwise, the opponent may draw first. In this case, the Sucker's payoff—which comes from letting the other guy draw first—is so poor, and the Temptation so high, that each participant may well feel induced to "defect" and, thus, to draw and fire. The tragic outcome is for the two gunslingers to shoot at each other and for both to receive the Punishment (P) of mutual defection:

		Gunfighter 2	
		cooperates (doesn't shoot)	defects (shoots first)
Gunfighter I	cooperates (doesn't shoot)	R, R	S, T
	defects (shoots first)	T, S	P, P

Both gunfighters would be better off, on balance, if they shook hands, clapped each other on the back, and sauntered off to the local saloon to buy each other a round of drinks. But that didn't happen very often.

Thus far, we have considered only Prisoner's Dilemmas that are strictly symmetrical, in which the options of the two players are interchangeable. This needn't necessarily be the case; a 2×2 game situation is a Prisoner's Dilemma so long as the payoffs are arranged in the relationship $T > R > P > S$, even if one player's Temptation, for example, isn't exactly the same as another's.

Take our next example, from the opera *Tosca*. In this case the Prisoner's Dilemma component was first pointed out by game theorist Anatol Rapoport. Scarpia, the evil police chief, has arrested Cavaradossi, lover of the beautiful Tosca. Cavaradossi has been sentenced to death, and Scarpia, that scoundrel, wants sex with Tosca. So he makes her this offer: If Tosca will give herself to him, Scarpia will order that Cavaradossi's firing squad be issued blanks so Tosca's beloved will not be executed after all. Now, the game begins:

Tosca sees herself as having two options. She could have sex with Scarpia ("cooperate" with him), or "defect" in an especially lethal way: pretend to comply/cooperate, but when Scarpia embraces her, stab him! This is particularly tempting to Tosca, since not only does it enable her to save her sexual honor, it also—she thinks—gets her true love returned to her.

Scarpia, however, sees things differently: He can "cooperate" with Tosca and carry out his part of the deal, or he can "defect," in an especially lethal way . . . for Cavaradossi. Thus, if he secretly countermands his promise to Tosca and has the firing squad use bullets instead of blanks, Scarpia reasons that he will not only get Tosca but also do away with his rival, Cavaradossi, whom he detests. The Tosca game looks like this:

| | | Scarpia | |
		cooperates (saves Cavaradossi)	defects (kills Cavaradossi)
Tosca	cooperates (succumbs to Scarpia)	R, R	S, T
	defects (kills Scarpia)	T, S	P, P

For Tosca, the best outcome would have been for her to kill Scarpia—who really *is* a swine!—while the deceived police chief kept his word and saved Cavaradossi (T). The worst would have been for her lover to lose his life while she succumbed to Scarpia (S). For Scarpia, the best outcome is sex with Tosca while Cavaradossi dies (T), and the worst is to be killed while Cavaradossi survives (S).

"I can resist anything," Oscar Wilde once announced, "but temptation." In Puccini's opera, both Tosca and Scarpia give in to their personal Temptations, and both defect: Tosca kills Scarpia, only to discover that Scarpia has, in fact, arranged for the firing squad to use real bullets after all! (Tosca then jumps out a window when the police come for her.)

The double-cross is, regrettably, the most logical resolution of a Prisoner's Dilemma. Indeed, it is the only logical resolution! Imagine this: A kidnapper has abducted the beloved daughter of a wealthy family and wants to exchange her for a certain amount of money, say $1 million. The victim's family has the money and is equally eager to make the exchange. Add this complication: The money must be yielded up in one place, while the victim is to be returned somewhere else, simultaneously. That is, kidnapper and family must each make their decision independently of the other. The family and the kidnapper are each afraid of being suckered: The family fears that they may drop off the $1 million and still not get their daughter back; moreover, they may also be tempted by the possibility of getting her back without having to pay anything. Similarly, the kidnapper fears that he may release his victim and not get paid; moreover, he may also be tempted by the possibility of getting the money while not releasing the girl, whereupon he can make yet another demand for more money. (Or—shades of the *Tosca* game—maybe he is also tempted to kill the victim and thereby eliminate a witness who might later identify him to the police.)

		Kidnapper	
		cooperates (gives up victim)	defects (keeps or kills victim)
Family	cooperates (gives the money)	R, R	S, T
	defects (uses fake money)	T, S	P, P

Following the now-familiar Prisoner's Dilemma matrix, both the victim's family and the kidnapper may therefore be induced to defect. The family leaves a bag filled with blank paper, and the kidnapper does not release the girl. Each has double-crossed the other. As a result, the kidnapper still has no money, and the family does not have its daughter. Both would have preferred an exchange—girl for money—but each was fearful of being suckered as well as tempted to "stick it to" the other, and, as a result, both lost out.

There are, sadly, all too many situations in which everyone loses out when a Prisoner's Dilemma is resolved in favor of mutual defection. Take political campaigns, alluded to in chapter 1. To go negative or not to go negative: That is the question most politicians face when things get serious and, often, desperate. To switch from Prisoner's Dilemma to politician's dilemma, substitute "go negative" for "defect" and "stick to the issues" for "cooperate." Instead of receiving jail terms of different lengths (although this has been known to happen!), politicians get such payoffs as "winning," "losing," "being seen as a nasty person," and so forth. Contending politicians would likely be better off if neither goes negative, analogous to both prisoners spending only one year in jail, or the kidnapper getting his ransom and the family getting their daughter back, or—more parallel, in a sense—neither gunslinger drawing first.

But the evidence is clear that even though no one likes negative campaigning, it does convey an electoral advantage to the perpetrator—that is, the one who "defects"—especially if the victim does not retaliate in kind. And so both candidates feel a logical pressure to go negative: like our hypothetical prisoner, who fears that he might be played for a Sucker while at the same time hoping to take advantage of his opponent, our very real candidates typically end up going negative. And as with the prisoners, everyone is worse off as a consequence. The public gets less reasoned debate, and each candidate gets the onus of being seen as a negative, nasty campaigner, maybe even a dreadful person. Who comes out ahead? Only political consultants and those television and radio stations that run those negative ads.

Or imagine that two individuals are arguing about something, almost anything. If their dispute involves the possible exchange of money—as recompense for damages, for example—each would almost certainly be better off if represented by a lawyer, especially if the other does *not* retain counsel. The frequent result: Both sides hire attorneys, and, as a consequence, neither is better off. In fact, both are probably worse off, since on average everyone hires an equally competent attorney, who must be paid!

Don't think that Prisoner's Dilemma applies only to individuals, however, or even that it must be hurtful to everyone. Consider two businesses, for example, gas stations on opposite sides of an intersection. The owners are best off if they cooperate and keep their prices high. But each is tempted to reach for a higher payoff yet: to defect and sell gasoline at a slightly lower price so as to get more business at the other's expense. And possibly to drive the other out of business. The result is—or can be—a ruinous price war that hurts the owners but benefits the public. (At least, that's the theory; in practice, companies have many ways to collude, to conspire in restraint of trade, thereby minimizing their vulnerability to Prisoner's Dilemma. As we'll see a bit later, there are similar possibilities for a kind of beneficent collusion by which the rest of us can solve our own Prisoner's Dilemmas . . . or at least grant ourselves a fighting chance to overcome them.)

Countries, too, often find themselves prisoners of dilemmas of their own making. Take arms races. In the admittedly simplified world of game theory, a pair of adversaries—think of the United States and the Soviet Union during the Cold War, or India and Pakistan in the early twenty-first century—have two options, call them "disarm" (cooperate) versus "arm" (defect). If both cooperate, then both receive the payoff R, a fairly high Reward for their budgetary restraint, since they are able as a result to direct their resources to more productive sectors of their economies, and presumably improve the lives of their citizens. If both defect, on the other hand, then both receive the punishing payoff P, which results from mutual wasting of national treasure. But each is sorely tempted to defect nonetheless, because of the seductive hope that by doing so, the defector might receive T, the payoff of an uncontested military buildup, which presumably allows its possessor to dictate terms to the "sucker" who elected to disarm and finds itself faced with a more powerful opponent.

The onset of war often resembles a game of Prisoner's Dilemma (although, as we'll see in chapter 5, it is equally likely to follow the pattern of another game—and one that is even more troublesome—known as Chicken). Take World War I, frequently considered the classic case in which ostensibly sophisticated, well-meaning diplomats blundered into something that neither side wanted, as though they were prisoners of a pernicious logic beyond their control. Tensions had been high for some time, as England and Germany in particular had been engaged in an escalating arms race:

> *In England they have built two dreadnoughts more*
> *The House of Lords is stormily debating*
> *The press screams, "Germany is plotting war!"*
> *While all of Europe buys up armor-plating.*[3]

Then, in midsummer of 1914, at a crucial point in the fateful run-up to that disastrous encounter, the heir to the Austrian throne is assassinated by a Serbian extremist, and the Austrian Empire, feeling itself threatened, issues an ultimatum to Serbia that is tantamount to a demand for surrender. Germany is allied to Austria and has to decide whether to support its Teutonic brethren (and risk war with Serbia's Slavic big brother, Russia, as well as its Western Allies, England and France), or to back down, dissociate itself from the war-mongering, and cooperate with the Western Allies, thereby seeking a diplomatic solution but risking humiliation. For their part, the Allies (England, France, and Russia) have to decide whether to oppose the German and Austrian Central Powers—and risk war—or to back down; that is, to defect or to cooperate, to risk war or to be more accommodating and pursue peace (but with the chance that they might end up being humiliated).

The decision matrix may have looked like this:

		Germany decides to	
		back down (cooperate)	support Austria (defect)
Allies decide to	back down (cooperate)	Agreement, Agreement	Humiliation, Prestige
	oppose Germany (defect)	Prestige, Humiliation	War, War

Neither side knew whether the other would defect or cooperate. Each placed great value on the prestige to be gained by defecting, and/or feared the humiliation that might come with backing down. Hence, the Temptation to defect was very high, just as the fear of being a Sucker was great. At the same time, although both sides may well have recognized the merits of reaching a peaceful, cooperative agreement (R, the Reward for mutual cooperation), each feared seeming weak, and so the Allies decided to oppose Germany while Germany decided to support Austria; together, the great powers of Europe blundered, in a kind of lockstep, into the mutual Punishment of war.

Looking at this slightly differently: The Allies didn't know if Germany would support Austria or back down. They knew, however, that either way, they had two choices: oppose the Central Powers or back down, and in a sense, their internal monologue was very much like that of the hypothetical prisoner stuck in his classic dilemma. "If Germany supports Austria, and I back down, I am humiliated; if Germany supports Austria and I oppose Germany, the result will be war, but I prefer it to humiliation. On the other hand, if Germany backs down in the face of my vigorous opposition, I gain enormous prestige; if Germany backs down and so do I, the result is an agreement. I prefer prestige. So, whatever Germany does, I am going to prepare to go to war (that is, defect)." With the German leadership reasoning similarly, the result was mutual defection.

Another case of Prisoner's Dilemma between countries: trade barriers. Nations would probably be better off, on balance, if they all experienced free trade. But each is benefited, individually, by erecting tariffs against the other. So they tend to do so. The problem is that if either country unilaterally permitted the other to trade freely on its soil but didn't get similar treatment, the unilateralist would be worse off. The result is a Prisoner's Dilemma in which each exacts tariffs against the other, and, as a result, both are worse off. Significantly, the only way out of this appears to be international agreements that prohibit tariffs, with penalties in the event of defection.

Looking dispassionately at all this, it is easy to forswear rationality altogether. After all, participants in a Prisoner's Dilemma are prisoners of nothing so much as their own reasoning. If they could only free themselves of the stranglehold of their own overactive cerebral hemispheres, they might be free to "do the right thing" for everyone concerned. Maybe so, but turning off one's own reason is a bit like taking a medicine that only works if, when you swallow it, you do *not* think of elephants! In addition, there is growing evidence that even creatures altogether lacking a brain cannot evade the clutches of Prisoner's Dilemma.

For example, an article appearing in the prestigious journal *Nature* described Prisoner's Dilemma in—of all things—an RNA virus.[4] Here's how it works: This particular virus, known as a phage, infects bacteria. There are two different kinds of this phage, both frequently found together inside a single bacterium: the "normal" form, P-6, and a mutant variant, P-H2. P-6 is an honest cooperator in that once inside a bacterial victim it manufactures some of the products needed for its own replication. P-H2, on the other hand, is a selfish defector, since it makes fewer of these necessary nutrients, relying when possible on P-6, which has to do double duty as a result. So, when P-6 (the cooperator and potential Sucker) is present, P-H2 gets something of a free ride. When only P-H2 is present, these selfish phages suffer the cost of mutual defection, since there is no other virus to take advantage of.

Here is a payoff matrix, based on rounded-off actual values determined for the reproductive rates of the two different kinds of virus, depending on whom they are sharing a bacterium with:

	P-6 (cooperator)	P-H2 (defector)
P-6 (cooperator)	1.0 , 1.0	0.7 , 2.0
P-H2 (defector)	2.0 , 0.7	0.8 , 0.8

Evidently, the problem of Prisoner's Dilemma is much older than game theory. But do not despair. There appear to be solutions, some of them new, some nearly as old as viruses themselves.

Forward Thinking and Backward Induction

You may have already noticed one problem when it comes to applying Prisoner's Dilemma to the real world. The various examples of Prisoner's Dilemma—gunslinger, *Tosca,* kidnapper, not to mention the classic case of two forlorn and manipulated prisoners—are all one-shot affairs. The two players encounter each other just once, after which they presumably never meet again. Under such conditions, it is not surprising that defection would be especially attractive, even irresistible.

But what about those cases in which Prisoner's Dilemma Games are repeated (in official terminology, "iterated")? When each player knows that there will be ongoing interactions and, thus, the prospect of benefiting from a bright future of positive, cooperative encounters—or, alternatively, the looming danger of continuing Punishment in the future—wouldn't that overturn a shortsighted tendency for selfish defection? Putting it more technically, even though logic dictates defection in one-time-only games of Prisoner's Dilemma, wouldn't the beckoning prospect of future reciprocations, studded with a string of potentially rewarding payoffs, induce the participants to cooperate? Unfortunately, the answer is no.

To understand why, consider this: Let's say you anticipate one hundred games of Prisoner's Dilemma. The dilemma, as we have seen, is that if you and your partner/opponent cooperated each time, instead of defecting, both of you would walk away with a higher payoff . . . but in any given game, it is terribly tempting to defect, nonetheless. Imagine that you and the other player have somehow managed to restrain yourselves through the first ninety-nine games, perhaps because despite the Temptation, neither of you wanted to provoke the other into defecting on the next move. You have both been happily pocketing the payoff R instead of P, and now, you are engaged in the last encounter, game 100. What should you do? The answer is clear: You should defect! Since there will be no game 101, there is no reason to be nice on game 100, and every reason to try at last for T, which is, after all, higher than R. Moreover, you may well worry that your partner/opponent is no different from yourself. Hence, you suspect that he, too, is inclined to try for T . . . which is all the more reason for you to defect!*

So, your best move is to defect in the last game, 100, regardless of what the other player does. Since you would do that, what is your best move in the game immediately before that one; namely, game 99? Once the future is foreclosed by the need to defect on game 100, then game 99 becomes, in effect, the last game, and by the same logic that caused you to defect in game 100, you should defect in game 99, too. And so on, back to game 1.

---

*Actually, a player's subjective probability that the other will cooperate is somewhat beside the point because no matter *what* the other does, it pays to defect.

According to legend, the great astronomer Arthur Eddington had just delivered a masterful lecture on the structure of the solar system, whereupon someone from the audience objected that it was all nonsense: everyone knew that the Earth was balanced on the back of a giant elephant. "And what," asked Eddington, "does the elephant rest upon?" "A giant turtle." "And what does the turtle rest upon?" "You're very clever, young man, but you can't get out of it that way. It's turtles all the way down!"

The distressing result is that no matter how many future games you may have to look forward to, the relentless logic of Prisoner's Dilemma impels you—and the other player—to defect. The future doesn't change a thing. It's turtles all the way down. (By the way, the same logic applies to defection among animals, even though our furred, feathered, and scaly friends—including turtles—are presumably unable to figure out the nasty logic just presented. The key point is that evolution will have worked it out for them, conferring a reproductive advantage on their ancestors who defected first, unswayed by the siren call of potential interactions in the future.)

Lest you get too depressed by this finding, I should point out that there appears to be a way out, nonetheless. But first, let's also clarify the process just outlined, by introducing a more cheerful game. Call it "Say Twenty First." This game—and the simple trick for winning it—involves a logical process known as "backward induction," the same phenomenon that causes iterated games of Prisoner's Dilemma to be logically no different from single games.

In "Say Twenty First," the first player says either "one" or "one, two." Then the second player adds one or two sequential numbers. So if the first started with "one," the second player can say "two" or "two, three." And so on. (If the first player started with "one, two," the second player can then say "three" or "three, four.") The goal in "Say Twenty First" is—not surprisingly—to be the one to say "twenty" first. Think about it: Whoever says "seventeen" wins, since if the opponent then says "eighteen," the other player can say "nineteen, twenty," whereas if the opponent says "eighteen, nineteen," the other player can say "twenty." So the way to win "Say Twenty First" is actually to be the one to say "seventeen," which means—using the same logic—that the goal is to say "fourteen" . . . which is achieved by saying "eleven" and, if necessary, before that, "eight," "five," and "two." So, by backward induction, we arrive at the conclusion that whoever goes first in this game can guarantee a win, by saying "one, two." (Or, if the other player starts, and says "one," you're sitting pretty if you say "two.")

"Say Twenty First" can be a cute little parlor game, until the trick is realized. Then, it's no longer worth playing, since the result is predetermined, as with tic-tac-toe, or, regrettably, iterated games of Prisoner's Dilemma.

What, then, is the way out? Just this: Although a repeated series of Prisoner's Dilemma Games does not offer escape from the dilemma, the logic of backward induction breaks down if the last game in the sequence remains unknown. Paradoxically, then, cooperation is irrational for a sequence known in advance to comprise, say, 367 games, but it could indeed develop if the sequence is shorter (or longer), so long as its precise termination point remains shrouded in uncertainty. So iterated games of Prisoner's Dilemma can, in fact, logically lead to cooperation so long as the future is sufficiently unclear!

And finally, there is yet another cause for optimism, wrapped—as is so often with game theory—in paradox. It turns out that people aren't as strictly logical as theory might predict (more on this in chapter 7). It is indeed true that strict logic calls for noncooperation in a Prisoner's Dilemma, whether playing just one round or if the final round is known. But experimental simulations have shown, time after time, that most people simply don't cooperate with such self-serving "rationality." Instead, they're inclined to cooperate with each other, to their mutual benefit.[5]

Why? It's hard to say. Probably the best explanation is also the most obvious: Just as the backward induction paradox isn't immediately obvious to most people, it is entirely likely that the perverse "wisdom" of mutual defection in a lengthy series of iterated Prisoner's Dilemmas isn't either. If so, then such unawareness must be seen as a kind of blissful ignorance, one that benefits both players . . . so long as both are equally benighted!

Some people may also perceive (whether acutely or just in dim outline) that cooperation in iterated games of Prisoner's Dilemma, although strictly illogical, can also generate dividends. Imagine, for example, that you are able to play a series of one hundred iterated games with someone else with Rewards provided by an external "bank," such that your payoff for cooperating on a given round is $1 (if both of you cooperate), and for defecting (if the other cooperates), $1.10. Imagine, also, that if you cooperate when your partner defects, you lose 10 cents; if you both defect, no one gets anything:

		Your partner	
		cooperates	defects
You	cooperate	$1, $1	−10c, $1.10
	defect	$1.10, −10c	0, 0

In this situation, if you both cooperate, you each stand to win $100 by the end of the game ($1 per play), even though the strictly logical thing is to defect each time. If you did that, both you and your partner/opponent/fellow prisoner will have earned nothing. But, of course, you will have avoided the possible worst case loss of $10: a hundred rounds at 10 cents per round if you kept cooperating while the other guy kept defecting. So, maybe there is a kind of underlying rationality at work. Maybe the turtles are resting on something solid after all.

For Anatol Rapoport, there are two kinds of rationality,

> *individual* rationality, which prescribes to each player the course of action most advantageous to him under the circumstances, and *collective* rationality, which prescribes a course of action to both players simultaneously. It turns out that if both act in accordance with collective rationality, then *each* player is better off than he would have been had each acted in accordance with individual rationality.[6]

For another, similar perspective, look at it this way: Imagine that both players understand the downside of being stuck with the Punishment of mutual defection, and, accordingly, they use backward induction to see how they would be better off cooperating. The result has been called a "nonmyopic equilibrium," in which both players can benefit from each other's farsightedness.[7] (But then, what if each is more farsighted yet and, perceiving that the other will cooperate, is fatally tempted to defect? By being so "farsighted," one risks a dangerously myopic outcome.)

Other Ways Out?

At first blush, there seem to be numerous problems with Prisoner's Dilemma, aside from its unrealistic focus on one-time interactions. We'll look briefly at a few.

For one thing, what about the restriction on communication? Remember that in the classic formulation of Prisoner's Dilemma, the two prisoners are kept in separate cells, unable to communicate with each other. The idea is that the clever district attorney is thus able to keep them from influencing each other, and, as a result, each is likely to imagine the worst about the other's intentions. (Recall Earl Silbert's manipulation of John Dean, getting him to think that Gordon Liddy was going to make a deal, thereby inducing Dean to defect first.)

As a general principle, there is no doubt that more communication is better than less—at least for the communicators, although presumably not for the district attorney. Economist Robert Frank reports that in experiments he conducted—in which small amounts of money were the payoffs in simple games of Prisoner's Dilemma—subjects turned out to be 60 to 75 percent accurate at predicting whether someone else would defect or cooperate if they were given a brief time (10 minutes to a half hour) to interact with them, in pairs, before playing.[8] This suggests that people are pretty good at "reading" each other, which, in turn, suggests that cooperativeness could prosper, even in a competitive, Prisoner's Dilemma–type environment, especially if cooperators might be able to seek each other out and benefit from interacting with each other, instead of with nasty would-be defectors.

But unfortunately, it isn't so simple. Let's go back to our two prisoners and imagine that they were kept in the same cell and allowed not only to assess each other, but also to cajole, threaten, beg, and bargain to their hearts' content, before entering a plea with the D.A. What then?

It seems logical that each player wants to convince the other that he will cooperate. (You certainly don't want to convince the other that you will defect, because in that case, the only logical response is for him to do the same.) So, we can imagine that each player will swear on a stack of Bibles that defection is the furthest thing from his mind, that he gives his word of honor, guaranteed, double-pinky promise, to cooperate. But there is a big, ugly fly in this ointment: What is to stop the most persuasive communicator, after promising up, down, and sideways to cooperate, from defecting when push comes to shove? And why wouldn't the other player be aware of this risk? In short, so long as the payoffs remain as indicated in the 2 × 2 Prisoner's Dilemma matrix, it doesn't matter what either player says; the potential of what he might *do* remains unchanged. This is especially true if the consequences are severe, as in a real Prisoner's Dilemma, with real prisoners. Similarly, it's no coincidence that in the realm of military, strategic analysis, where game theory has been especially influential, a careful distinction is drawn between a potential opponent's *intentions* and its *capabilities*. When national security is at stake, walking the walk is more important than talking the talk.

Moreover, to make matters worse, if either player seems especially believable in his protestations of cooperation, such that the other is convinced of his goodwill and honest intentions, wouldn't this merely tempt the seeming cooperator all the more strongly to defect at the last minute, and thereby take maximum advantage of the situation? Moreover, adding additional layers of complexity, what is to stop the naïve recipient of such protestations—especially if he isn't all that naïve— from figuring that this is precisely what the would-be cooperator is trying to induce, which in turn would make the recipient of such protestations all the more wary and, thus, even *more* likely to defect?

Communication, in short, isn't what it's cracked up to be. At least not in games of Prisoner's Dilemma.

What about attacking the underlying premise of the game, that players will try to get the highest payoff possible? Maybe there really is honor among thieves. (Alternatively, how about just among the rest of us?) Going back to the generalized Prisoner's Dilemma matrix, it is noteworthy that three of the four designated payoffs had distinctly theological overtones: Temptation, Reward, and Punishment. Sucker stood out like a sore thumb. It can be corrected, however, and this goes to the heart of the "honor among thieves" critique. Instead of Sucker, let's call the lowest payoff—the one received by a cooperator when faced with a defector—Saint. Maybe his Reward will come in heaven, or perhaps for the Saint, doing good is its own reward: "Take that, game theory! Some people are motivated by things above and beyond your grubby payoffs!" If so, and if the Saint is moved by such considerations outside the ken of the Prisoner's Dilemma matrix, then the logic of the Prisoner's Dilemma itself wouldn't apply. Or, to put it differently, what if people do things for reasons that don't involve maximizing their payoff?

There is nothing wrong with this thinking, except for this: It's incorrect. Not that people aren't moved by "other" considerations, such as doing what's right for its own sake, saving one's soul, being noble and virtuous because that's what you were taught, or placing great value on genuine altruism. (It is interesting, by the way, to consider that martyrs and true altruists are pretty rare, probably because left to their own devices most people are inclined to maximize their payoff in the here and now; hence, the persistent efforts of moralists—religious and secular—to convince people to "do the right thing.") The key point, however, is *not* that people sometimes behave in seemingly unselfish ways, not readily amenable to calculation. Rather it is this: Insofar as people are influenced by such factors, this doesn't in any way defeat a Prisoner's Dilemma. It doesn't move the situation into a different realm, in which payoffs are irrelevant. Once again, it simply changes the value of the payoffs in question.

Thus, for a Saint who is perfectly willing to suffer for the equivalent of turning her other cheek when someone else is tempted to slap it, the bottom line isn't that she relishes a low payoff and thus defies logical analysis, but rather that, for her, turning the other cheek and possibly being slapped offers a higher payoff yet! What is really required is simply that we take additional payoff considerations into account. The Saint or honor-among-thieves approach "solves" the Prisoner's Dilemma . . . but not really, because all it does is transform it into some other game.

This is an answer, of sorts. And it may be the best one available: The way to solve a Prisoner's Dilemma is to change it to something else.

Along these lines, some say that the way to satisfy a Prisoner's Dilemma is to come to an enforceable agreement, with the government, for example, requiring cooperation. A posse of sheriff's deputies can intervene in the gunfighter's shootout, pointing their rifles at the two desperadoes and insisting that both of them leave their revolvers holstered; that is, "cooperation" by constraint. An international commission can enforce free-trade zones. Ransom/prisoner exchanges can be supervised by a powerful but neutral third party. Dominant male baboons will sometimes act as peacekeepers among squabbling females, and adult geese (of both sexes) occasionally intervene to keep order among goslings. All this is wonderful, and, in practice, it can solve many dilemmas. But neither enforcement nor resort to "hidden" or "higher" payoffs solves the general problem, since if enforcement exists or the payoffs are actually different, then once again, the particular impasse is no longer a Prisoner's Dilemma. To repeat, maybe that's the answer: Solve Prisoner's Dilemmas by stripping them of their Prisoner's Dilemma-ness.

The preceding touches upon one of the fundamental issues in ethics and political philosophy: How to get cooperation among egoists in a world lacking an overpowering central authority? If individuals—or companies, or countries—are simply out for their own gain, how can they be induced to behave in a mutually helpful manner, short of being forced to do so? The great German philosopher Immanuel Kant had an answer. He called it the "categorical imperative," and it is essentially a variant of the Golden Rule. Kant said that when it came to evaluating the suitability of any particular action, we should ask ourselves whether we would be satisfied if our choice became a "general rule," adopted by everyone else.

If we are tempted to lie, cheat, or steal, we must ask ourselves if we would accept a world in which everyone else did the same. From then on, according to Kant, reason would be our unfailing guide. If adhered to, Kant's categorical imperative is a way out of the dilemma. But it's easier said than done, largely because it simply tells people to cooperate when the payoffs are in the form of what we recognize today as a

Prisoner's Dilemma. As such, it identifies the problem and points toward Anatol Rapoport's "collective rationality." If everyone followed the Golden Rule—"Do unto others as you would have them do unto you" (Luke 6:31)—the Prisoner's Dilemma would dissolve.

But, sadly, the categorical imperative doesn't really eliminate the dilemma. The problem is that pure rationality points, if anywhere, more strongly in the other direction. It is altogether rational—although in the long run, highly destructive—for individuals to proclaim adherence to the categorical imperative (that is, announce one's intention to cooperate), but to defect nonetheless! Kant himself would doubtless object at this point, however, pointing out that such defection would itself fail his criterion for a categorical imperative.

According to Thomas Hobbes, writing in the mid-seventeenth century, people simply can't be trusted to pursue the common good when tempted with its alternative: the opportunity for private wealth or power. Personal greed, for Hobbes, will always trump civic virtue. Left to their own devices, most people will defect most of the time. Accordingly, a strong government is absolutely necessary, precisely because of the human resistance to cooperating. Living in a state of nature, Hobbes famously claimed, human existence would be dominated by a selfish penchant for what game theorists identify as defection, and, as a result, human lives would be "solitary, poor, nasty, brutish, and short." Hobbes's answer was "Leviathan," the heavy-handed Guardian State whose powerful central authority would keep all of us fellow prisoners in line, thereby resolving the great dilemma.

Modern biologists understand that Hobbes was unduly pessimistic about the "state of nature," at least insofar as its nonhuman inhabitants are concerned. Ants and termites, for example, live industrious and more or less harmonious existences in vast hives, wolf packs maintain a laudable degree of social propriety, and herons nest by the hundreds in remarkably close proximity without much happening that qualifies as nasty or brutish. Yet, one could argue that these and other animals have already internalized their own species-specific version of Leviathan, via their unique, instinctive tendencies for social control, cohesion, and communication. And when it comes to our own species, it is difficult, after witnessing the violent chaos that followed the demise of Saddam Hussein's regime in Iraq, to avoid the conclusion that people are liable to behave very unpleasantly indeed if left to their own devices.

Hobbes's solution is eminently practical, no matter how unpleasant. It is also unsatisfying, at least for the game theoretician, since, by introducing a powerful policing agent, Hobbes makes the cost of defecting so high that it is no longer an alluring Temptation.

But for our purposes, the problem remains an intellectual challenge no less than a practical matter: How to get out of a *genuine* Prisoner's Dilemma? (Without making it into something else, and without calling in the 101st Airborne.)

Axelrod's Tournament and Rapoport's TIT-FOR-TAT

Robert Axelrod, a professor of political science at the University of Michigan, had a brilliant idea. He would contact some of the best minds in game theory and ask each to submit a recommended strategy for how best to play a simple iterated game of Prisoner's Dilemma. After all, since Prisoner's Dilemma seemed to encapsulate something terribly important—and frustrating—about the problem of cooperation, why not look seriously for a solution? But how to determine the best way out? Axelrod figured that it might be useful to introduce a variety of Prisoner's Dilemma strategies in a round-robin computer tournament, and let them duke it out, checking at the end to see which strategy amasses the highest overall payoff. He therefore asked fourteen different specialists in game theory—from five disciplines: economics, mathematics, political science, psychology, and sociology—to contribute his or her recommended strategy, in the form of computer code.

Each entry consisted of a different "rule," a program that coded for a particular strategy of cooperation or defection during a repeated game. For example, a very simple program might be "always defect," or "always cooperate." A more complex one could consist of something like "defect every third game," while the most sophisticated take the other player's previous behavior into account, resulting in strategies such as "defect whenever the other player has cooperated twice in a row," or "cooperate if the other player defects; defect when the other player cooperates," or even highly sophisticated rules that attempt to use previous responses to assess the probability that its opponent/partner will cooperate after experiencing either a defection or a cooperation from itself.

As a round-robin tournament, each entry was paired with every other, with each game consisting of two hundred repetitions. (Each entry was also paired with itself and with a random program.) The payoff matrix for the tournament was a basic Prisoner's Dilemma, as follows, showing a confrontation between two computer programs, 1 and 2:

		Program 2	
		cooperates	defects
Program 1	cooperates	3, 3 (R, R)	0, 5 (S, T)
	defects	5, 0 (T, S)	1, 1 (P, P)

You can see that T, the Temptation to defect, offers 5 units of payoff; R, the Reward for mutual cooperation, offers 3; P, the Punishment for mutual defection, is 1; and S, the Sucker's payoff—received by a cooperator paired with a defector—is 0. Each program accumulated points as the games proceeded, and at the end, the programs were ranked in order of score.

The winner—submitted by the doughty Anatol Rapoport, was called TIT-FOR-TAT. It is the simplest program of all the entrants, and it called for the following strategy: Cooperate on the first move; from then on, do whatever the other player did on the previous round. In other words, if the other player cooperates on move 1, TIT-FOR-TAT will cooperate on move 2. If the other player defects eventually, let's say on move 193, TIT-FOR-TAT will defect on move 194; if the other player goes back to cooperating on move 195, TIT-FOR-TAT dutifully resumes cooperation at the next opportunity, move 196.

In his superb book *The Evolution of Cooperation,* published in 1984, which describes this tournament and its implications, Axelrod noted that for all the emphasis on competition, defection, and sneaky aggressive strategies in games of Prisoner's Dilemma,

> surprisingly, there is a single property which distinguishes the relatively high-scoring entries from the relatively low-scoring entries. This is the property of being *nice,* which is to say never being the first to defect.

It also became clear that the outcome was highly dependent on the available entries, the various strategies against which each one played. So Axelrod decided to conduct a second round. This time, all invited participants were given extensive information as to the outcome of round one, and everyone knew that all the others had received the same information. There were sixty-two entries from six countries in round two. Rapoport once again submitted TIT-FOR-TAT, unchanged from round one, and once again, it came in first. This is a striking result, because of the insight that can be gained from TIT-FOR-TAT itself. (More of this to come.) It is also remarkable that even though TIT-FOR-TAT was known to have been the initial winner, no one could come up with anything better!

Before going further, let's be clear that TIT-FOR-TAT is not necessarily "the" solution to Prisoner's Dilemma. The fact that it emerged victorious in both rounds of Axelrod's tournament could be a consequence of the payoff structure (5, 3, 1, and 0), or of the "strategic environment" (the other programs with/against which it performed). Or something else. Nonetheless, it is likely that TIT-FOR-TAT has much to teach us. Let's look briefly at some of these lessons.

For one, being nice—that is, not being the first to defect—consistently paid off. For another, being too nice isn't a good idea either. Doormats get walked upon. In Axelrod's terminology, it pays to be "provocable." Because TIT-FOR-TAT does whatever the other player did on the previous move, it cannot be taken advantage of. Or at least, not for long. If the other player defects, TIT-FOR-TAT responds on the next move, so it cannot be suckered very often. In his book, Axelrod acknowledged that one of the surprises for him was the importance of "provocability." He began this research, he noted, believing that it was a good thing to be slow to anger, but now feels that prompt provocability (combined with equally prompt willingness to forgive) is very helpful.

At the same time, TIT-FOR-TAT is indeed "forgiving." That is, unlike some programs that responded to a single defection with an unending string of punishing defections, TIT-FOR-TAT is quick to resume cooperating as soon as the other player does the same. It doesn't hold grudges.

As it happens, an even more forgiving strategy than TIT-FOR-TAT—namely TIT-FOR-TWO-TATS, in which defection only occurs after *two* defections by the other player—would have done better yet. In fact, it would have won the first tournament!* Moreover, this strategy was used as the sample mailed to each contestant, demonstrating how to make a submission. It is striking that among the chosen game theory experts, not one submitted TIT-FOR-TWO-TATS. Maybe this is because each was so egotistical as to think that he could do better than the sample. But as Axelrod suggests, it is also possible that "even expert strategists do not give sufficient weight to the importance of forgiveness." (Or, perhaps this is an error that "expert strategists" are *especially* likely to make. There may be something to be said for naïveté.)

Axelrod summarized the simple secrets of TIT-FOR-TAT's success by pointing to

> its combination of being nice, retaliatory, forgiving, and clear. Its niceness prevents it from getting into unnecessary trouble. Its retaliation discourages the other side from persisting whenever defection is tried. Its forgiveness helps restore mutual cooperation. And its clarity makes it intelligible to the other player, thereby eliciting long-term cooperation.

Most interesting, perhaps, is this paradox: TIT-FOR-TAT never defeated its opponent. This is so striking that it deserves to be repeated. TIT-FOR-TAT emerged as the overall winner, even though it never received a higher score than the other player in any series of games. To

---

*On the other hand, it did rather poorly in Axelrod's second tournament, because several other competitors were prepared to take advantage of it.

see how this works, consider that in order for a player to get a higher score than its opponent, it must defect when the opponent cooperates. This cannot happen on the first move, since TIT-FOR-TAT begins by cooperating. After that, it might well defect, but only if the opponent had defected just before, in which case it simply balances the scales (and, moreover, only some of these retaliatory defections by TIT-FOR-TAT encounter a cooperative move from the other side). In the worst case, TIT-FOR-TAT is defeated by one defection: after being suckered a single time, it ties even the nastiest opponent from then on. TIT-FOR-TAT achieves its high overall score by evoking cooperation from the other side, after which both sides get the moderately high payoff, R. At the same time, TIT-FOR-TAT avoids being consistently suckered because it effectively punishes any defections, while repeatedly giving the other side the opportunity to repent from any previous defections. The malefactor is punished, but quickly, cleanly, and without long-term consequences.

By contrast, highly defection-prone programs did on occasion defeat their opponents, particularly when paired with others that were excessively forgiving. But when highly aggressive programs encountered each other, the outcome wasn't pretty: Each got caught up in a string of retaliatory defections so that they both ended up with low, punishing payoffs (mutual P). TIT-FOR-TAT, meanwhile, just kept moving along, defending itself against meanies while rewarding kindlies, and, of course, rewarding itself at the same time (via the comparatively high payoff, R). "Joint undertakings stand a better chance," we learn from the ancient Greek playwright Euripedes, "when they benefit both sides."[9]

So why doesn't everyone engage in mutually beneficial "joint undertakings" all the time? Don't forget the fundamental dilemma: Even though an individual can often do well by cooperating, it remains true that each will do better yet if it could exploit the cooperative efforts of others. The marvelous quality of TIT-FOR-TAT is that although it doesn't exploit, it doesn't get exploited. And it rewards cooperation so that, in fact, both sides ultimately stand to benefit.

TIT-FOR-TAT isn't meek, however. As we have seen, because of its provocability, it falls short of the Christian ideal. But then again, so does just about everyone. On the ethical plus side, TIT-FOR-TAT is nice (it initiates cooperation), it is restrained (it doesn't take advantage when the other side behaves cooperatively; TIT-FOR-TAT responds to cooperation by cooperating, rather than being tempted to defect), and it is also forgiving (if the other side switches from defecting to cooperating, TIT-FOR-TAT follows suit, never holding a grudge). It changes Prisoner's Dilemma from a competition into a collaboration. These virtues were enough to win Axelrod's computer tournament. They might even enable their practitioners, someday, to inherit the earth.

Before going on, however, let's introduce a bit more reality, which may be frustrating for anyone inclined to see TIT-FOR-TAT as offering God—or even, secular salvation—in a computer algorithm. The bottom line: TIT-FOR-TAT is wonderful, and illuminating, but it isn't *the* solution to Prisoner's Dilemma. Why not? First, because it isn't necessarily the best—highest payoff—way to play. Rather, it was the strategy that came out on top in Robert Axelrod's two computer tournaments. This doesn't mean that it is guaranteed to be the number one world champ. As already mentioned, other strategies (including TIT-FOR-TWO-TATS) would have done even better. It is also possible to design different competitions in which TIT-FOR-TAT wouldn't be such a success, such as an elimination tournament instead of a round-robin.[10]

In addition, TIT-FOR-TAT is vulnerable, not just to certain other strategies that are especially nasty, but also to old-fashioned mistakes. If you introduce into a TIT-FOR-TAT program the reasonable assumption that every once in a while, a player defects—simply by error—and allow these programs to play with/against each other, they quickly get mired in an unending series of mutual defections. Game theorists call it the "trembling hand" problem, when even the best of strategies occasionally blunders. A key quality (more easily demanded than achieved) is for a strategy to be stable against this inevitable sling and arrow of outrageous fortune: when you reach for an ideal strategy, only to miss, because your hand happens to shake.[11]

It happens, all too often. The Israeli Defense Forces, seeking to "eliminate" a Hamas or Islamic Jihad terrorist leader, end up killing innocent Palestinians, whose compatriots respond by murdering Israeli citizens at random, whereupon the Israeli government authorizes yet more assassinations of Palestinians. Similar "TIT-FOR-TAT killings" have long plagued Northern Ireland, the Balkans, and elsewhere. In short, once we add the complexities of the real world—notably the inevitability of mistakes—TIT-FOR-TAT loses a bit of its luster.

Hence, in the world we actually inhabit—one filled with what engineers call "noise," by which they are not referring to decibels—TIT-FOR-TAT needs to be tinctured with both generosity and contrition. Exactly how to accomplish this is a task for theorists and well-meaning policy makers alike.[12]

At the same time, just as TIT-FOR-TAT may have been excessively hailed as the game theoretic savior of cooperative niceness, it is also too easy to disparage it simply because it isn't all things to all people. Thus, a series of computer simulations have shown, impressively, that TIT-FOR-TAT is an especially viable strategy even when Prisoner's Dilemma is played in a spatial environment as well as a temporal one.[13] And theorists have demonstrated that social pressure, including ostracism of noncooperators, can strongly predispose players toward "doing the right thing." After all, if defectors are identified and excluded from successive rounds, they lose the possibility of Temptation, Reward, even Punishment; they don't get to play at all, which, for a highly social and interactive creature such as *Homo sapiens,* may be the greatest Punishment imaginable.[14]

Most important, TIT-FOR-TAT shows that Prisoner's Dilemma does not necessarily condemn humanity to the bleak prospect of unending defection and failed cooperation, even if some pattern other than TIT-FOR-TAT turns out to be better yet. Axelrod's work opens the door to the importance of continued interaction—lengthening the "shadow of the future"—showing that the rewards of cooperation can indeed be genuine, and not simply a longed-for utopia populated by well-meaning simpletons.

The "Live-and-Let-Live" System During World War I

Robert Axelrod described an especially striking example of TIT-FOR-TAT in action; namely the "live-and-let-live" system that developed on the Western front during World War I. It is so dramatic and constitutes such a remarkable confirmation of TIT-FOR-TAT that it deserves to be recounted here. Actually, "live-and-let-live" never existed as a formal "system." Rather, like the slave girl Topsy, in *Uncle Tom's Cabin*, it "just grewed," in this case out of the horror of trench warfare, emerging by quiet common consent among the participants. Game theoreticians had nothing to do with live-and-let-live; the theory wouldn't be developed for several decades. But in retrospect, we can see that reality preceded theory . . . which is, after all, the way these things are supposed to work.

When ordered to do so, troops on either side of the World War I trenches would attack one another, with horrible effect. But between the large set-piece battles, soldiers typically refrained from doing serious harm to their counterparts, so long as such restraint was reciprocated. In short, even in the midst of a notoriously murderous war, Germans on one side and French and British on the other readily cooperated with one another, so long as those across the trenches did the same. "Let sleeping dogs lie" would be another way to describe the system, or perhaps, "If it ain't broke, don't fix it." Maybe it was actually closer to "Things are already bad enough; don't make them worse." However you slice it, the basic idea is straightforward enough: For the individual solider in the trenches, although the cost of being suckered—shot by a sniper—was high, there was little to be gained by shooting at each other (the Temptation to defect was low). Accordingly, individuals on both sides were inclined to reap the Rewards of mutual cooperation; that is, to refrain from endangering one another.

This live-and-let-live system was not the same as the famous informal Christmas Truces that developed on December 25, 1914 and 1915, when soldiers briefly and openly fraternized across the infamous "no-man's-land," in at least one case even sharing an informal soccer scrimmage. The Christmas Truces were isolated incidents. Live-and-let-live, by contrast, became—literally—a way of life for prolonged stretches of time.

Axelrod points out that in practice, the strategy employed wasn't always TIT-FOR-TAT. Often it was two for one. (Professor Axelrod refrained from using the obvious label, but I am not so delicate: We're talking here about TWO-TITS-FOR-A-TAT.) A doubly severe response of this sort is not surprising: Since the cost of being suckered was so high, it makes sense that the system was biased to punish any defection, promptly and severely. In short, it was provocable. Consider this memoir of a British officer:

> It was the French practice to "let sleeping dogs lie" when in a quiet sector . . . and making this clear by retorting vigorously only when challenged. In one sector which we took over from them they explained to me that they had practically a code which the enemy well understood: they fired two shots for every one that came over, but never fired first.[15]

The higher command, on both sides, hated live-and-let-live, and did everything possible to disrupt it. At least one British staff officer, touring the trenches, reported with indignation that he was astonished to observe German soldiers walking about within rifle range behind their own line:

> Our men appeared to take no notice. I privately made up my mind to do away with that sort of thing when we took over; such things should not be allowed. These people evidently did not know there was a war on. Both sides apparently believed in the policy of "live and let live."[16]

This informal "policy" was admirably recounted by sociologist Tony Ashworth, in his book, *Trench Warfare, 1914–1918: The Live and Let Live System.*[17] Ashworth shows that the system was not—as some self-serving British accounts suggest—found only among the French soldiers, nor was it—as some French military diarists maintain—a peculiarly British institution. Ashworth's account also makes it clear that live-and-let-live was not an isolated peculiarity; he draws from numerous sources, showing that the system flourished within nearly every one of the fifty-seven British divisions, with suggestive evidence that it was comparably widespread among French and German soldiers as well.

There was no need for the "players" to be friends, or even to know each other. Cooperation based on certain fundamental principles evidently can develop even between two sides that have vowed to destroy each other.

History students typically learn how the leading diplomats of Europe found themselves captured by the narrow expectations of military treaties, face-saving, and even rigid railroad schedules, as the continent lurched toward "the war that no one wanted." For their part, the soldiers in the trenches were also prisoners in a military and political dilemma not of their making. Paired units—typically a battalion on each side—facing each other across a few hundred yards of no-man's-land were players in an iterated game of Prisoner's Dilemma. For them, more than for the diplomats and senior officers, the consequences of overweening aggressiveness threatened to be immediately lethal; accordingly, recourse to patterns of potentially life-saving cooperation (actually, restraint), underwritten by rules of TIT-FOR-TAT reciprocity, became a pressing concern. Newly arrived recruits were quickly indoctrinated with the rules of the road, as with this recollection: "Mr. Bosch ain't a bad fellow. You leave 'im alone; 'e'll leave you alone."[18]

If either side violated such tacit understandings, the response was typically prompt and unambiguous:

> The real reason for the quietness of some sections of the line was that neither side had any intention of advancing in that particular district. . . . If the British shelled the Germans, the Germans replied, and the damage was equal: if the Germans bombed an advanced piece of trench and killed five Englishmen, an answering fusillade killed five Germans.[19]

Since units consisting of the same individuals typically faced each other for prolonged periods of time, a dilemma persisted, on both sides: It was prudent to shoot to kill, since there was an advantage to weakening the enemy on the other side of the line in the event that a battle would subsequently be ordered. Thus, T was greater than R. But mutual restraint, R, was greater than P, the Punishment of shooting and getting shot at. And similarly, exchanging fire, P, was better than being shot at without giving any response (S). The traditional outcome in a one-time Prisoner's Dilemma—in which both sides defect and both receive the Punishment of their mutual defection—is what the commanding officers on both sides wanted. But this didn't serve the lower-level soldiers, the individual prisoners.

Early in World War I, battles were mobile and quite bloody. It is significant that mutual restraint and nonaggression developed only later, once the lines stabilized and trench warfare began. Periods of especially bad weather, or regular times of resupply on both sides of the trenches, became opportunities around which mutual cooperation could develop and cohere, like crystals forming around specks of dust. Altruism wasn't the point; instead, there was intuitive recognition that restraint served the interests of the participants on both sides, much as military forces have long been expected to avoid attacks on hospital units, or Red Cross facilities.

The inhabitants of the trenches, on both sides, needed resupply. A soldier noted, accordingly, during 1915 that

> It would be child's play to shell the road behind the enemy's trenches, crowded as it must be with ration wagons and water carts, into a blood-stained wilderness . . . but on the whole there is silence. After all, if you prevent your enemy from drawing his rations, his remedy is simple: he will prevent you from drawing yours.[20]

It is interesting to note that one source of tension, threatening to undermine the live-and-let-live system, was the fact that artillery and infantry were not equally vulnerable to retaliation, and, thus, they did not have an equal stake in maintaining the system. Artillerymen were more insulated and, therefore, more willing to defect; infantrymen, as a result, often found themselves seeking to diminish the former's aggressive ardor. On occasion, they even apologized for it, as seen in the following incident, related by a British officer:

> I was having tea with A Company when we heard a lot of shouting and went out to investigate. We found our men and the Germans standing on their respective parapets. Suddenly a salvo arrived but did no damage. Naturally both sides got down and our men started swearing at the Germans, when all at once a brave German got on to his parapet and shouted out "We are very sorry about that; we hope no one was hurt. It is not our fault, it is that damned Prussian artillery!"[21]

Ashworth reports that when artillery units sent new forward observers to the front lines, these "spotters" were often greeted with the following message from their own infantry: "I hope you aren't going to start trouble." Similarly—although more subtly—with newly arrived officers, especially midranking recent graduates of the various service academies, typically patriotic and well schooled in military theory, but unfamiliar with the realities of how to survive in the trenches.

But of course, the High Commands on both sides hated anything that smacked of a tacit truce or live-and-let-live. They were in the war to win it—or at least, to receive decorations for making a credible effort in that direction—not merely to survive. They undermined the live-and-let-live system by court-martialing anyone who obviously refrained from shooting to kill (despite which, incidentally, soldiers on both sides became adept at shooting in a manner that avoided retaliation). Officers then began disrupting the informal "system" by ordering destructive raids on each other's lines, which, true to the expectations of TIT-FOR-TAT, evoked a violent response. In fact, this may well have been the real purpose of such raids: not so much to damage the other side as to keep the pot of mutual defection boiling . . . or rather, to keep it from congealing into what the officers considered an unacceptable stew of mutual cooperation.

Even more successful was the eventual strategy of rotating units so as to break up any possible pattern of mutual cooperation among battalions facing each other. It takes time for each "player" to establish a degree of trust in the other's restraint, combined with demonstration of provocability. And for mutual cooperation to develop, it is helpful—indeed, necessary—for both participants to know that each is likely to be around to play the next round. Otherwise, the old dilemma rears its ugly head, and defection beckons.

When analyzing the results of his computer tournament, Axelrod pointed repeatedly to "the shadow of the future," emphasizing that for TIT-FOR-TAT to be effective, it is helpful if this shadow is long, so that both players see themselves as having an investment in maintaining an ongoing, mutually beneficial relationship. This is why seasoned trench soldiers hated being shipped to a new sector almost as much as they hated their own new officers. It is also part of the reason why live-and-let-live didn't carry the day, although it provides a dramatic case of TIT-FOR-TAT in the all-too-real world, and a poignant example of people struggling to free themselves from an especially lethal dilemma.

The sad reality is that TIT-FOR-TAT, for all its potential as a means of extricating prisoners from their awful dilemmas, carries a terrible danger: that its provocability can trigger comparable behavior in the other side, whereupon both players find themselves mired in a swamp of retaliation. This is not simply a matter for intricate and occasionally amusing laboratory or computer simulations . . . although it has generated more than its share in this regard. Rather, it is often splashed with pain and blood, the daily lacerations of mutually inflicted human misery.

Here is Israeli novelist David Grossman, writing in the *New York Times* after an especially dreary and destructive chain of TIT-FOR-TAT violence between Israelis and Palestinians:

> Both sides are in a mad, dizzy spiral of violence. The lunatic logic of this conflict holds that if you have not responded with full force to the blow you suffered, the other side will interpret it as weakness and will strike at you again even more painfully. The result is that each side is doomed to hit its antagonist, and then it cringes in fear of the counterblow.
>
> The rhythm of life, the rhythm of consciousness, all contact between one person and another—everything is conducted entirely according to the tick of this deadly metronome. In such an atmosphere, who even remembers that the real goal that we must aspire to is not the next attack on the enemy, or effective protection against him, but to attempt to bring this cycle of death to an end?[22]

Reciprocity

The key element in TIT-FOR-TAT is, not surprisingly, a key element in much human interaction: reciprocity. It goes by many phrases, such as "fair exchange," "you scratch my back, I'll scratch yours," or—on an avowedly positive side—"that's what friends are for" as well as—on the negative—"don't tread on me."

People often exchange small favors: Christmas cards, running helpful errands for a neighbor, inviting someone to dinner. So long as both sides cooperate, the system is typically stable, even though either party is free to defect. And, in fact, if one "player" begins defecting, the likelihood is that eventually the other will do the same, in which case both receive P, the Punishment for mutual defection. In other words, the relationship is finished or, at least, at risk. If George and Martha invite John and Abby to their house for dinner, the expectation is that next time, it will be John and Abby's "turn." If, as time goes on, they don't meet this expectation, the likelihood is that the two couples will cease to be close friends. Their relationship may survive one or two failures of reciprocation, but beyond that it becomes pretty clear that one side—in this case, John and Abby—are sending a message. (Of course, this isn't necessarily a genuine loss for either side, and such termination is often intended by the nonreciprocating party, in which case mutual defection isn't all that punishing, for one side or both. The point is simply that expectations of reciprocity are so clear that any failure to abide is readily interpreted.)

The system is quite sensitive. There even appears to be a kind of built-in warning mechanism that lets us know if we are violating the expected norms of reciprocity, perhaps by accident. Most people feel guilty, for example, if they receive a Christmas card from someone who isn't on their list. It feels wrong to have someone cooperate, in a sense, with you, when you aren't doing your part. In fact, the basic notion of exchange—whether barter, purchase, or the simple concept of "you help me, I'll help you"—seems fundamental to human society. Our species is even equipped with brain regions specialized for recognizing the faces of others, and it is at least possible that face recognition is itself crucial to reciprocating, TIT-FOR-TAT systems. Thus, for mutual cooperation to be stable over time, recipients must be able to identify who cooperated with them in the past, and who is therefore worthy of reciprocal cooperation, and also who defected and may therefore warrant a defection in return. (Recall the "live-and-let-live" system of World War I . . . and, of course, its lethal retaliatory counterpart, which was essentially "shoot and get shot at.")

An interesting thing about exchanges of mutual cooperation is that most of them take place a little bit at a time, with potential Rewards and Punishments parceled out in small packages, very much like a series of repeated games of Prisoner's Dilemma. There is wisdom here, since as we have seen, cooperation and benevolent reciprocity (albeit of the self-serving sort) is more likely in a large number of small games than just one big one, in which the Temptation to defect is liable to be much greater. Interestingly, this pattern of low-consequence, repeated games, with careful monitoring of the other's response, takes place among animals as well.

For example, biologist Manfred Milinski set up a nifty experiment using stickleback fish, a common freshwater inhabitant of northern lakes and streams. When not breeding, sticklebacks live in schools, and are, of course, vulnerable to larger, predatory fish. When a large, hungry-looking fish shows up, several sticklebacks will make an "inspection visit," in which they approach the potentially dangerous predator. Each inspector is at risk, but is somewhat protected by the behavior of the others; so long as all advance together, they divide the risk. No one wants to be left hanging out alone. In this situation, inspector fish advance in a series of small, jerky approaches. It seems likely that each is monitoring the actions of the others, making each step a kind of iterated Prisoner's Dilemma. (To go toward the possible predator is to cooperate; to hang back is to defect.)

Milinski installed a series of mirrors, as a result of which a single fish was deceived into thinking it was accompanied by others. Actually, it was seeing its own reflection, from different angles. The way the experiment was set up, each subject stickleback's mirror image could be made to seem that it was cooperating (approaching when the test animal approached), or defecting (turning tail, literally). The results, in short: Nearly all test animals followed TIT-FOR-TAT. First they cooperated (approached). Then they did whatever their mirror did: approached if it approached, fled if it fled.[23]

Animal behaviorist Eric Fischer studied another TIT-FOR-TAT system, which, coincidentally, also involved fish. The subjects were coral reef dwellers known as black hamlets, aptly named in a sense, because they reveal a kind of "to be or not to be" uncertainty. The question is whether to be male or female, and these particular hamlets resolve the question by choosing both. They are "simultaneous hermaphrodites," that is, they produce both eggs and sperm. Since eggs and sperm are shed at different times, each piscine hamlet still needs a mating partner. We need to realize at this point that eggs are more expensive to produce than are sperm, and, so, a hamlet would be more fit (produce more descendants, in the long run), if it could assume the male role—that is, use its abundant and cheap sperm to fertilize the eggs of several other hamlets—rather than be relegated to femalehood, which means allowing others to employ their cheap sperm in fertilizing "her" expensive eggs. Fischer found that mating among these animals involves a sophisticated pattern of "egg trading," in which the sexual partners take turns being male (ejaculating sperm) and female (releasing eggs), in a repeated series of four to twelve spawning acts.[24]

Finally, consider another case of an iterated Prisoner's Dilemma, once more in an aquatic setting: skinny-dipping. (This example is admittedly more fanciful than the previous two, but personal experience may well confirm its general outline.) Some people may be turned on by exhibiting their naked bodies to the view of others, but in most cases, the secret little pleasure of taking off one's clothes in the company of others is seeing *them* nude. Most skinny-dippers aren't seeking to be the only one undressed, with everyone else leering and making fun, or in some other way benefiting from one's exposure. Taking off an article of clothing is thus an act of cooperation; remaining clothed while others are stripping is a defection. So clothes come off determinedly, but with little furtive glances all around, not so much to see another's body as to make sure that everyone else is doing the same. What sticklebacks do when inspecting a predator—that is, make sure their colleagues are cooperating and not defecting—skinny-dipping human beings do when inspecting each other.

Future Tripping

We've seen that one-shot games of Prisoner's Dilemma are much more likely to produce defection than are repeated games. This is because the potential benefit of mutual cooperation the next time can sometimes act as a positive lure, inducing both players to "play nicely" so long as the future looms bright. The more it is cut off—the less light at the end of the tunnel—the greater the inclination to beat each other up (or defend against being beaten up), even if the outcome is benighted indeed, for both parties.

As mentioned, Axelrod employed a memorable phrase, noting that cooperation is more likely when "the shadow of the future is long." On the other hand, when this prospect is limited, TIT-FOR-TAT is less attractive, and so is cooperation, generally. Julius Caesar offered this explanation of why allies of his rival, Pompey, ceased cooperating with him once Pompey's situation became very weak: "They regarded his prospects as hopeless and acted according to the common rule by which a man's friends become his enemies in adversity."[25]

Maybe it is going too far to say that in adversity, one's friends turn into enemies, but certainly fair-weather friends are thereby revealed. Similarly, a lame-duck president can expect less support than one who is continuing in office, or who has at least the potential of being reelected. This is why politicians are typically loath to announce their retirement, until it can no longer be avoided, since political power shrinks dramatically as soon as it becomes evident that one is no longer available for future deal making, whether positive cooperation or the threat of punishing another's defection.

Here is an example from business. When a firm is on the edge of bankruptcy, its accounts receivable are often sold to an outside party known as a "factor." When this happens, the factor purchases the right to collect on these accounts at substantially less than their official value. Such discounting is nearly inevitable because

> once a manufacturer begins to go under, even his best customers begin refusing payment for merchandise, claiming defects in quality, failure to meet specifications, tardy delivery, or what-have-you. The great enforcer of morality in commerce is the continuing relationship, the belief that one will have to do business again with this customer, or this supplier, and when a failing company loses this automatic enforcer, not even a strong-arm factor is likely to find a substitute.[26]

By the same token, relationships are more likely to be trusting when they are keyed to long-term interactions, whereas most people would be reluctant to purchase an insurance policy, for example, from someone who is just "passing through" (unless the company in question has an especially solid, long-standing reputation). It is no coincidence that Harold Hill, the slick-talking salesman-hero of *The Music Man,* was portrayed as a conscience-less predator, who made his living by bilking credulous customers whom he expected never to see again. Similarly, the word "gyp," a critical slang expression meaning "to swindle or cheat" derives from Gypsy. It isn't clear whether real Gypsies—or, as they call themselves, Romany—are actually more likely to engage in unreciprocating, exploitative business practices (that is, to defect), but it is easy to see how their traditionally nomadic lifestyle would have contributed to such a perception.

Stereotypes of this sort are often unfair, but we can nonetheless begin to glimpse how the tendency to generalize in this way may have originated. Society enforces cooperation, in part, by labeling individuals, associating them with their names, addresses, social security numbers, and so forth. (And perhaps in the near future, retinal images or DNA fingerprints.) Labels, for all their pernicious potential, connect people with their earlier behavior. Once labeled, we can't long avoid the consequences of our past actions. If we are defectors, we can be associated with that fact; similarly if we have proven to be reliable cooperators. Human beings doubtless evolved in small groups, within which we were certainly known. For better or worse, our ancestors couldn't escape developing a reputation. With the growth of large urban units, it is at least possible to become anonymous, and with such anonymity, there may well be a greater likelihood of defecting. And this, in turn, has probably given added impetus for the current penchant for labeling that so bedevils those worried about personal privacy.

A kind of labeling occurs among animals, too, and maybe even plants. As a result, cooperation can still evolve, even when brains are so small that individuals may be unable to recognize a partner (not to mention look up a credit report), or when there is simply so much social traffic that the prospects of the same two individuals encountering each other again some time in the future may be low. In such cases, the trick is to make the association long-term, if not permanent: anemone fish are therefore likely to live with their particular sea anemone; termites have a long-standing relationship with the symbiotic gut-dwelling bacteria that enable them to digest wood; trees and their root fungi are intimately and cooperatively interconnected, and so forth.

Alternatively, if lifestyle requires that the "players" go about their separate existences, they may establish a fixed meeting place, as with cleaner fishes and their hosts. Thus, part of the natural economy of many oceanic fishes calls for big lunkers to be cleaned of their parasites by small cleaner fish that swim, with impunity, inside their "customers'" gills and even mouths. Both cleaner and cleaned benefit, but since the larger fish could easily eat the smaller, and the smaller could also refuse to service the larger, it is important for both "players" to establish reliable meeting places and recognition behaviors . . . which they do. Coral reefs are dotted with "cleaning stations," where cooperative exchanges regularly take place between trusted, old-time partners. Once again, in such cases, it may be tempting for either party to defect: for the cleaner fish to take a chunk out of the larger animal being cleaned, or for the latter to swallow the former. But in the long run, it pays both parties to refrain and to cooperate instead.

Prisoner's Dilemma as Rorschach Test

Early in the twentieth century, Swiss psychiatrist Hermann Rorschach pioneered the test that now bears his name. The idea was—and still is—that people reveal their mental traits by telling what emotion or story is evoked in them by a series of inkblots. Different people see different things in a Rorschach test. By the same token, different people respond in different ways to games of Prisoner's Dilemma. Maybe they see different things in the same situation. Certainly, we can see things in *them* depending on how they respond (and also, how we choose to interpret each response, which may say as much about us as about them!).

Here are a few possibilities; in each case, C = cooperate and D = defect. Therefore, CC = mutual cooperation; DD = mutual defection; CD = you cooperate and the other player defects; DC = you defect and the other player cooperates.

1. If you play C after having experienced CC in the preceding exchange, then you are trusting, trustworthy, hopeful (or maybe boring and unimaginative).
2. If you play C after having experienced CD, then you are forgiving, a martyr, a saint (or maybe a naïve fool).
3. If you play C after having experienced DD, then you are sadder but wiser (or maybe a reformed sinner, or just unrealistically optimistic).
4. If you play C after having experienced DC, then you are guilt-ridden, repentant (or maybe eager to cover up your transgression and hope no one notices).
5. If you play D after having experienced DC, then you are greedy or triumphant (or maybe inclined to teach the other party a lesson . . . for his own good, of course!).
6. If you play D after having experienced DD, then you are angry, self-protective, defensive, resentful, spiteful (or maybe just self-protective).
7. If you play D after having experienced CC, then you are manipulative, sneaky, selfish (or maybe just unwilling to be stereotyped as a "goody two-shoes"). And finally,
8. If you play D after having experienced CD, then you are jaded, cynical (or maybe just sadder but wiser).

Here is an interesting finding, easier to describe than to interpret: When individuals are called upon to play Prisoner's Dilemma just once—so that the only rational choice is to defect—about 40 percent choose to cooperate.[27] Admittedly, it is troublesome that 60 percent choose to defect. But this also means that fully 40 percent cooperate, which is likely to be encouraging to those who value cooperation over defection as an ethical ideal. At the same time, maybe the finding that 40 percent of subjects cooperate when doing so is altogether irrational is itself cause for *discouragement,* on the part of those people who value rationality above all else! Once again, the Rorschach seems to be on us.

Another example: Most people interpret the decision to cooperate as a positive, benevolent overture, but it could also be seen as a sneaky, underhanded effort to lure the other into cooperating, so that one can then reap the benefit of defecting next time! Maybe the real motivation for cooperating is the hope of inducing one's opponents to lower their guard, all the better to exploit them, my dear. On the other hand—considering the long-term benefits that come from being part of a system of mutual Reward—maybe cooperation is simply a matter of enlightened self-interest, and thus not really altruistic or especially ethical at all.

Similarly, defection is typically seen as offensive, selfish, and exploitative, whereas, actually, it may be defensive, motivated simply by fear of the other rather than hope for ill-gotten gain at the other's expense. This may explain a peculiar, but consistent finding: Compared to men, women regularly exhibit *lower* levels of cooperation when playing staged games of Prisoner's Dilemma. This violates most people's expectation based on traditional gender roles, since men are typically more competitive and women more relationship-oriented and, presumably, more cooperative as well. Yet, one classic study found 60 percent cooperation in male pairs versus only 34 percent in female pairs.[28] It is at least possible that insofar as women are often exploited in other contexts, they are generally more defensively oriented; expecting the worst, perhaps, they are inclined to "defend" themselves, albeit preemptively. (But even this explanation, reasonable as it may seem, must be at least somewhat lacking, since the same sex difference arises in the Game of Chicken, where—as we'll see in chapter 5—defecting is hardly the safest alternative. The effect of sex differences on cooperation/defection is currently unresolved.)

In any event, it is interesting that when people are interviewed as to whether they expect someone else to cooperate in staged games of Prisoner's Dilemma, cooperators—whether men or women—are far more likely to anticipate cooperation from others. And *vice versa*. In one representative study, defectors predicted a rate of 24 percent cooperation from other players, whereas cooperators predicted cooperation of 68 percent.[29] This makes perfect sense, because a major reason for defection is the expectation that the other player will defect. At the same time, it may simply reveal people trying to justify their own nasty, selfish behavior by suggesting that "other people do it, too."

Neither martyrs nor cynics do very well in Prisoner's Dilemma: Martyrs are often Suckers and likely to be exploited mercilessly. Cynics assume the worst from their colleagues, and, as a result, that's exactly what they bring about.

Thus, in another study, subjects were asked to describe their own inclinations (whether competitive or cooperative) before playing forty rounds of Prisoner's Dilemma. Every ten rounds of play, the subjects were then asked to evaluate the intentions and style of their opponents. Those subjects who were, by their own account, cooperative were able to evaluate the strategy and style of their opponents with good accuracy after only ten games. By contrast, competitive subjects were much less accurate; in particular, they were unable to identify the other players' cooperative intentions when such intentions existed.[30] A likely explanation: Whereas cooperative people experience both cooperative and competitive "partners," competitive people are likely to evoke competitiveness in others. There is a German saying that captures this: *Wie man hineinruft, so schallt es heraus.* ("How one shouts determines the echo.") This is especially true of that particular kind of ethical echo chamber known as Prisoner's Dilemma.

This leads to another view of Prisoner's Dilemma: as the crucible of paranoia. Admittedly, paranoid schizophrenics suffer from a severe psychosis, evidently due to neurochemical imbalances. But there is also a kind of "everyday paranoia," in which people are especially prone to feeling put-upon, taken advantage of, the butt of other people's animosity and ill will. This, in turn, generates a "paranoid stance," in which the self-identified victim believes he is being appropriately defensive, although others may well interpret his actions as aggressive. If you defect because you anticipate that others will do so, and then keep defecting because, in fact, others *are* doing so (perhaps because they are playing TIT-FOR-TAT), then you have unintentionally produced just the attacked/attacking relationship that is embodied in paranoia: a self-fulfilling prophecy. After all, as we have seen, TIT-FOR-TAT is notably forgiving as well as "nice." But TIT-FOR-TAT is also provocable.

"Even paranoids have enemies," goes the saying. Less frequently recognized is the extent to which paranoids—by expecting defection from others and therefore being defectors themselves—*produce* these enemies whom they so fear.

Also at issue is the degree to which Prisoner's Dilemma becomes a way of life, so that individuals—or companies, or countries—that have long been engaged in a Prisoner's Dilemma find it extremely difficult to extricate themselves . . . not so much from any particular Prisoner's Dilemma itself, but from the mind-set that such interactions produce.

Interestingly, people who are in favor of world government are more likely to cooperate in Prisoner's Dilemma; those who favor isolationism and who distrust international treaties are less cooperative.[31] It may well be that one problem the United States has encountered in seeking to adjust to the post–Cold War world is that we are still locked in the same Prisoner's Dilemma mind-set that characterized foreign relations for nearly a half century. Many Americans tend to see their options, even now, as constrained by another major and inimical player (Russia, China, or certain "rogue states"). As a result, stuck in a kind of residual Prisoner's Dilemma, we behave as though we cannot possibly afford to do the right thing (equivalent to "cooperate" or be "nice") out of fear that the other side will take advantage of us ("defect" or be "nasty"), leaving us with some sort of Sucker's payoff. But the world has changed, if, indeed, it ever was that way. Now, our major enemies are not another world power (or even, I would suggest, international terrorism) but environmental deterioration, disease epidemics, tidal waves of nationalism, and political unrest deriving from the increasingly grotesque economic imbalance between the "have" and "have-not" countries . . . precisely those kinds of opponents that cry out for a cooperative solution.

It is often difficult to realize the degree to which we make models of the world, often unconsciously, and then, the extent to which these models influence our actions. It happens to game theoreticians, in spades. Earlier, for example, I presented a nuclear arms race as an example of Prisoner's Dilemma. This is standard procedure among game theorists, who also, on occasion, use a different model: the Game of Chicken, which we shall explore in chapter 5. For now, let's note that although Prisoner's Dilemma and Chicken are different in one big way, they also share an important similarity: the assumption that the payoff is highest for whoever defects when the other cooperates. The Reward of mutual cooperation is ranked high, in both games, but lower than the Temptation to defect.

But what if this isn't the most accurate depiction? What if, instead, the highest payoff accrues to both players (in the case of an arms race, both countries) if they cooperate and refrain from escalating—or even playing—rather than to the side that builds weapons while the other refrains? Let's take a look at such a model for, say, a nuclear arms race between India and Pakistan, this time reworking it so that it is close to a Prisoner's Dilemma, but crucially different:

| | | Pakistan | |
		refrains	builds
India	refrains	4, 4	1, 3
	builds	3, 1	2, 2

In this case, both countries would be best off refraining—not building—even if building nukes meant that one would catch the other with its nuclear pants down. However, this configuration of payoffs still wouldn't guarantee mutual restraint, since both sides could worry that the other might behave irrationally and build anyhow, thereby getting a 3 and consigning the refrainer to a mere 1. (To be sure, 3 should be less desirable than 4, but what if either side is vulnerable to some sort of pressure from religious groups, or nationalistic passion, etc., that changes its subjective payoffs?) In such a case, both players need to be reassured that the other will not defect, that it will behave in a way that is ultimately in the best interest of both.

We might call these reassurance games, and they are quite realistic. In addition, fortunately, they are generally more easily solved than are games of Prisoner's Dilemma, since the key is simply to reassure each other rather than buck the nasty logic of the game itself. For all their real-life relevance, reassurance games are less intellectually challenging than Prisoner's Dilemma; actually, they are thinly disguised coordination games, in which both sides share a cooperative, mutual interest. In the above example, either side could encourage "transparency," allowing the other—or some international authority—to inspect its facilities, or simply pass a law or constitutional amendment making restraint mandatory.

The stumbling block in reassurance games isn't so much the game's inherent logic—after all, it is most logical to refrain, which is in everyone's best interest—but rather, a widespread tendency to distrust the other side. "Only the paranoid survive," claimed Intel chief executive Andy Grove, toward the end of the twentieth century. In the twenty-first, the survivors may well be those with the courage and vision to rise above paranoia.

Zero-Sum Jealousy

Here is a disturbing but important finding: In staged games of Prisoner's Dilemma, players often find themselves more concerned with reducing each other's payoff than with enhancing their own. Robert Axelrod, for example, says that he often has pairs of students play Prisoner's Dilemma in his class, asking them to play as if they were getting one dollar for each point. He also tells them to concentrate simply on accumulating as much money (as many points) as possible, and not to worry about how their partner is doing, or how they are doing by comparison. But inevitably, says Axelrod, individuals seek not only to enlarge their own payoff, but also to maximize the difference between their score and that of the others. The result is that those "others" change from partners to opponents.

Probably the best way to evaluate one's performance is not to compare it with the payoff received by some other player, but with what someone else would be doing if he or she were in your place and using a different strategy. But a kind of jealous one-on-one competition typically takes over when two players are interacting, so long as the payoffs are determined by the rules of Prisoner's Dilemma. Why? No one knows, precisely, but one likely explanation is that such interactions harken back to the kind of primitive competitive situations that may well have characterized much of early human evolution. And that still bedevil us today.

Remember that TIT-FOR-TAT—the winning strategy in Axelrod's renowned computer tournament—does not do better than any other player. In fact, it either ties the other, or does a little worse. The key to its victory is that it evokes responses from its partners that allow both players to do well, certainly better than they would have done otherwise: doing well by doing good. Is there something in the human psyche that keeps us from acting in our own enlightened self-interest, from doing as well (and as much good) as we should be capable of?

Here is a strange finding that speaks to this question. It is a situation set up decades ago by researchers at Ohio State University, a positive-sum game in which players on both sides should have had an easy time cooperating.

		Player 2	
		cooperates	defects
Player 1	cooperates	4c, 4c	1c, 3c
	defects	3c, 1c	0, 0

In this game (we might call it Cooperate, Stupid!) there is no reason for either player to defect. After all, everyone does best by cooperating, and the money comes from the experimenters not from the other player, so there are, in fact, some very good reasons *not* to defect. Admittedly, each side might be somewhat tempted to defect, hoping to catch the other cooperating; in this case, the defector gets 3 cents, the sucker only 1 cent. But each would get 4 cents by cooperating. Not only that, but if both defect, neither gets anything. Yet the researchers found that defection occurred, on average, nearly 50 percent of the time![32] Maybe this is simply an example of stupid irrationality. Or maybe it isn't all that irrational, since, as already suggested, much of life is, in fact, a zero-sum game (even if this particular game is not). Also, because the amount of money was so small, the players may have felt free to indulge their nastier inclinations, preferring their own satisfaction in suckering the other player to the actual amount that might otherwise be gained. On occasion, there may be more satisfaction in bettering the other player—or in avoiding being bested by her—than in actually getting the highest possible score. If so, then the actual payoff matrix would have to be rewritten, making it, in fact, a Prisoner's Dilemma.

There may be situations in which a shared payoff of +4, as in the Cooperate, Stupid! Game just described, is less conducive to social and biological success than one of +3 with the other side getting only +1. In particular, if players 1 and 2 are males competing for leadership in a small tribal band or, say, females competing over the success of their offspring, it may be that the key consideration for each player isn't so much the total payoff he or she garners, but the difference between his or her payoff and that received by the other player. In this case, we can rewrite the Cooperate, Stupid! Game as follows:

		Player 2	
		cooperates	defects
Player 1	cooperates	0, 0	−2, +2
	defects	+2, −2	0, 0

With this transformation, mutual cooperation is no longer so compelling. If player 2 cooperates, the best move for player 1 is to defect (+2 beats 0). And if player 2 defects, the best move for player 1, again, is to defect (0 beats −2). The same, of course, applies in mirror image to player 2. As a result, the best move for each player is to defect. This game, however, is not a Prisoner's Dilemma, because mutual defection doesn't generate a mutually punishing outcome when compared to the shared Reward of simultaneous cooperation. Bear in mind that this game was generated by assuming that the key isn't absolute payoff, but maximizing the difference between one's return and someone else's.

It isn't difficult, however, to imagine games in which cooperation should be stable. After all, despite the fact that competitive interactions—especially violent ones—receive a disproportionate amount of attention, the fact remains that there is a lot of cooperation going on in the world. So let's look at a theoretical payoff matrix in which the Rewards of cooperation clearly exceed those of "treachery."

		Player 2	
		cooperates	defects
Player 1	cooperates	50, 50	3, 3
	defects	3, 3	1, 1

Here, it does not pay to defect. Moreover, if you are already defecting, and whether you are player 1 or 2, it pays to switch to cooperation. So, cooperating should be the rule. Another way of looking at it: Offered these payoff prospects, it is difficult to imagine that even in a simulated laboratory situation, many players would do anything but cooperate. But even in cases of this general sort, it is possible to imagine situations in which rational defection may occur. Let's say that players 1 and 2 are not symmetrical; that is, they are starting from different initial conditions.

Imagine, for example, that a war is being fought, and "cooperate" means agree to a temporary truce, for both sides to bury the dead, or perhaps to obtain needed supplies. Although it might seem that the contenders would be equally inclined to jump at the opportunity to cooperate, since both would profit from such cooperation, this may not be the case. If one side has more food, more ammunition, or values burying its dead less than does the other, it may well be more resistant to a cease-fire and, thus, less willing to cooperate. It may value keeping the other player from getting its +50 more than it values obtaining its own +50, not out of spite but self-interest. (Once again, however, purists can point out that refusing to cooperate in this case wouldn't really be an exception to the logic of game theory payoffs, so much as it reflects a different pattern of payoffs for the two players.)

Finally, and on a similar note, consider the so-called Ultimatum Game (which we shall revisit in chapter 7). Two players have a pot of money, or perhaps a pie. Their job is to divide it up. The first player gets to propose the way it is divided, or to cut the pie. The second chooses. We already examined Cutting the Pie, as a coordination game, but in this case there is a catch: If the two players don't agree, no one gets anything. Logically, when playing an Ultimatum Game, if the first player suggests that he gets $99 and the second, only $1 (or if the person cutting the pie divides it into a very large chunk and a tiny slice, insisting that she will only accept the large one), the second should agree, despite the manifest unfairness, because otherwise she gets nothing.

But, in fact, when Ultimatum Games are actually played, participants routinely prefer getting nothing to seeing the other player walk off with an unfair advantage, even when the money involved represents several months' wages! This seems like incredibly bad judgment. And, in fact, it is, so that in some quarters the Ultimatum Game is gradually replacing the Prisoner's Dilemma as the poster child for human irrationality. But there could also be an underlying logic, reminiscent of the Cooperate, Stupid! Game, that goes something like this: If you accept the short end of the stick, and—perhaps worse yet—acquire a reputation for doing so, then you haven't only gotten a poor deal, you may also have set up your neighbor and competitor to be wealthier, stronger, and/or healthier than you. As a result, your neighbor may be more attractive to other women (or men) than you, better able to compete for other, more important prizes in the future. Add to this the risk that if your reputation suffers, others may be inclined to take advantage of you in the future, and it may be that refusing to agree to an unfair distribution—although narrowly illogical and downright spiteful—may turn out to make a kind of perverse "sense" after all.

Put this all together and, just maybe, closer attention to Prisoner's Dilemma will help all of us make greater sense of ourselves and our world.

4

Social Dilemmas: Personal Gain Versus Public Good

We're not yet done with Prisoner's Dilemma. In its traditional form, Prisoner's Dilemma depicts an interaction between equals, whether they be individuals, corporations, countries, or equivalent. More common, however, are situations in which an individual—or corporation, country, and so forth—plays against a larger collectivity: an individual against the herd or society, a corporation against the rest of the business community, a country against the world. You might think that in such cases, the loner loses. But not necessarily. In fact, because of the peculiar qualities of such situations, it is at least as common for the individual to come out ahead, at least in the short run, and for the group as a whole to be worse off as a result.

These cases have been called "social dilemmas," and they have a lot to teach us.

Some Examples

I teach a seminar-style course titled "Ideas of Human Nature." It's a good course or, at least, a popular one. For everyone's benefit, I restrict the enrollment to twenty students, because the special payoff of a seminar as opposed to a lecture is that students get to speak their minds and to hear others doing the same. This simply can't be done if the class is large (even twenty is a bit too big). But because the class is popular—which, ironically, is partly due to the benefits that come with keeping the class small—others want to get in. Most students understand the advantages of small classes of this sort, and so they wouldn't want everyone who wishes admittance to get in—just themselves. If everyone is admitted, the class becomes too big and discussion is inhibited, to everyone's disadvantage. And so, each year I find myself in the difficult position of telling a number of students that there simply isn't room for them. They can try again next year. Each student turned away from this class understands the logic, and even typically agrees with it, but nonetheless, each would like the limit to be expanded—by just one—to include him- or herself.

This is a social dilemma.

On the one hand, the class as a whole is somewhat worse off for every student above a given number (arbitrarily set in this case at twenty) who is admitted. On the other, each student who wants admission would be better off—or expects to be better off—as the twenty-first. Individuals seeking admission are playing "against" the rest of the class, and most are willing to "defect" by getting in, all the while assuming that the instructor will not allow *everyone* who wants entry to succeed. In fact, if I allowed unrestricted access to the class, it would lose its special value to all concerned.

If you are playing against the larger group in a similar situation, you, too, would probably be sorely tempted to defect—to choose the course of action that gave you a higher payoff—even if this somewhat diminished the payoff of the group as a whole, and even though you are also part of that group. The reason is that the downside for the group is generally distributed across many individuals, so the personal cost to the defector (in our seminar class example, someone who successfully "overloads" into the class) is likely to be low, whereas the benefit, for this person alone, is likely to be high enough that, on balance, he or she is better off being selfish.

Animal examples abound. Reproduction, in a strictly Darwinian perspective, is selfish. After all, baby making is the primary way living things project their genes into the future, thereby receiving an evolutionary payoff . . . actually, *the* evolutionary payoff. With this in mind, consider African elephants: Increasingly restricted to game parks, they end up fenced in and, as a result, locally overcrowded. Desperate to satisfy their huge appetites, hungry elephants strip the bark from trees, eventually killing them and, thus, ultimately destroying their habitat, to their own long-term detriment. But try explaining that to the elephants. For eons, natural selection has rewarded those who—selfishly and successfully—reproduced. Once again, it's a kind of social dilemma, whereby every elephant who becomes a parent gains an immediate evolutionary payoff, while in the process imposing a substantial long-term evolutionary cost on their habitat and, thus, on elephants as a whole.

Let's take a big leap in scale, to global warming. Instead of "add yourself to David Barash's class," or "become a mommy or daddy pachyderm," make it "add extra carbon dioxide to the earth's atmosphere." The accumulation of greenhouse gases in the atmosphere is having negative effects on the world's climate; this is obvious to all but industry lobbyists, a small number of contrarian scientists, and free market–worshiping government officials. Technical solutions exist; the real dilemma of greenhouse warming is a social one. It is often easier and, in the short run, cheaper to use polluting, carbon dioxide–spewing devices than to refrain, even though if everyone does it, we're all worse off. So, at the individual level, most people would prefer to drive their private automobiles rather than take public transportation, all the while bemoaning the resulting traffic, not to mention the ensuing buildup of carbon dioxide and its effects.

At the corporate level, firms are reluctant to cut back on their generation of greenhouse gases because it may place them at a competitive disadvantage. Each corporation is essentially inclined to defect, "playing against" other corporations, just as individual drivers are inclined to defect, "playing against" other commuters. At the national level, the United States under President George W. Bush refuses to abide by the Kyoto Treaty, complaining that it is not in the "national interest," even though it is clearly in the interest of the planet. (The United States, in this case, has elected to defect, "playing against" all other countries.) In such situations, the cost to the individual entity—person, corporation, country—of cooperating appears to be high, whereas the benefit seems low. Defection threatens to become the rule.

Which is precisely what happens in many cases.

There is a water shortage. Individuals should cooperate and conserve, but each is inclined to cheat. After all, many people like to have a lush, green lawn, or they may enjoy taking a lengthy shower, or they simply don't appreciate long delays between toilet flushes. People readily understand that in times of drought, it is important to conserve water. After all, the entire public—which includes themselves—would be in trouble if everyone else indulged a thirsty lawn, took long showers, or flushed often. But it is awfully tempting to cheat (that is, defect) because the cost of each private defection, to the group as a whole, is small, whereas the benefit to the cheating individual is potentially large . . . so long as only a few others do the same thing.

Or consider agricultural production. such as milk or corn. Paradoxically, farmers are better off when everyone produces less milk, say, or fewer bushels of corn, since this means higher wholesale prices. But at the same time, each is inclined to defect and produce more, since the marginal consequence on overall corn prices is trivial if Farmer Brown harvests a few additional bushels, while the benefit to Farmer Brown of "cheating" in this way and overproducing may be substantial; after all, he gets paid by the bushel, so the more bushels he produces, the more money he gets. He is therefore tempted to defect, all the while hoping that no one else does so.

A similar situation applies to OPEC members and oil prices. Or taxes: no one likes paying them. Indeed, people often go to great lengths to avoid doing so. At the same time, nearly everyone recognizes the benefits of living in a society in which people pay their taxes: it is beyond the capacities of individual citizens to maintain highways, libraries, fire departments, police forces, national defense, schools, hospitals, and so forth. And yet, it is awfully tempting to cheat on one's own taxes, or certainly, to err on the side of personal benefit . . . even though if no one paid his or her share, we'd all be in big trouble. Getting away with an unfairly small tax bill may constitute a major personal windfall, and in a country with predictable tax revenues in the trillions of dollars, the Temptation to defect is powerful indeed.

In fact, once you start identifying social dilemmas, it's difficult not to see them everywhere, whether in public affairs or private matters.

Take Cipro hoarding. During the autumn of 2001, widespread fears of anthrax bioterrorism induced many Americans—patriotic, well-meaning individuals—to defect from the public good and squirrel away their own supplies of Cipro, one of the few drugs proven effective against the disease. At the time, supplies of Cipro were none too abundant, and authorities urged citizens to refrain from hoarding it. But many did so nonetheless, fully recognizing that it would be a very bad thing if everyone did the same . . . although possibly a good thing if they were the only ones to "defect" in this way.

Other examples abound: ordering an expensive meal when it has been agreed that a group will divide the bill evenly; refusing to join a trade union and thus not having to pay dues, but profiting from whatever benefits the union is able to obtain for everyone, members and nonmembers alike; rushing for an exit during a theater fire (everyone's chances are better if each individual files out in an orderly manner, but you would probably get out more quickly if you pushed others aside); not giving to National Public Radio (but listening, nonetheless) or to your local blood bank (but accepting a transfusion if needed); or even something as trivial as standing on tiptoe during a parade (you'll get a better view, but others will be obstructed).

Finally, imagine that you are part of a group being flown to a remote fishing camp in Alaska. The bush pilot announces a total weight limit, above which the plane (and everyone in it) would be endangered. Clearly, no one wants this amount to be exceeded . . . and yet, you may well be tempted to bring along an extra few books, or perhaps your own bottle of Jack Daniel's.

Comparable Temptations are evidently felt by animals, too. Thus, biologists were intrigued when Robert Trivers, at the time a Harvard University graduate student, pointed out that "reciprocal altruism" can be expected among animals.[1] But despite more than three decades of serious effort to document self-sacrificing behavior among nonrelatives, remarkably few such examples have been discovered. We must conclude that the Temptation to defect is not uniquely human. Birds do it. Bees do it. Even monkeys in their trees do it. They give in to Temptation and typically refrain from giving to the group or, at least, from giving more than they have to. In fact, it may be that if anything, animals are *less* generous—more inclined to be selfish—than their human counterparts, because they don't have powerful ethical precepts (religion, morality, not to mention the IRS and the criminal courts) to remind them of their social obligations.

Mallard ducks, those lovely, iridescent green-headed denizens of freshwater ponds, provide an especially chilling example. Drake mallards are notorious rapists, which is to say, they force copulations with already mated ducks whose participation is obviously not consensual.[2] These sexual attacks frequently involve many males, and they are the mallard equivalent of gang rape among human beings. In the process, females suffer a high mortality because their heads are held under water during copulation, and, in the course of multiple, sequential rapes, they can drown. Why then, does male number two, and three, and so forth persist in forcing his sexual attentions on a female who may well die as a result? Clearly, this is a bad payoff for these males, and yet, each attacking drake is stuck in a social dilemma: to refrain, after other males have gone ahead, would be to increase the chance that the victimized female will survive, but with the guarantee that she will not be conceiving the "cooperator's" darling little ducklings. And so, male mallards "defect" and participate in gang rapes of females, to the detriment of the victimized females and even, to some extent, of themselves, stuck in a social dilemma of their own making, but one that is nonetheless devastatingly real.

Such self-generated realities may well become, if anything, more acute in the human future, too. Thus, with new techniques in genetic testing, it will become increasingly possible for parents to screen their unborn offspring, thereby avoiding a range of genetic diseases. There seems little doubt that most people, given the opportunity, would be inclined to do so, and not to be inhibited by any long-term cost to the diversity of the human gene pool. Similarly, the inevitable stresses of monogamy are somewhat ameliorated by the fact that the numbers of boys and girls—at least, at birth—are almost precisely equal. When and if techniques arise for parents to choose the sex of their offspring, and if current preferences for male offspring are maintained, it is possible that the sex ratio will eventually become unbalanced, and, along with it, some of what today passes for domestic harmony. (Actually, there is a powerful argument—itself based on game theory—showing that, eventually, sex ratios will settle down to 50:50; we'll encounter it in chapter 6.) But in the short term at least, don't expect the prospective parent, desperate to perpetuate the family name, to be moved by the fact that his preferences, multiplied several millionfold, might do harm to the larger social fabric. You could just as well try explaining to a mallard drake why he should be a gentleman instead of a rapist.

At root, these are all cases of Prisoner's Dilemma, the only difference being that in social dilemmas, each individual "plays" against a larger social group instead of one-on-one. If anyone needs convincing, here is a payoff matrix for the Water Shortage Social Dilemma; the same analysis—with only minor adjustment—would apply to all such cases:

		Everyone else	
		conserves water	waters lawn
You	conserve water	R, R	S, T
	water lawn	T, S	P, P

A Prisoner's Dilemma indeed: Your highest payoff (T, the Temptation to cheat) comes when you water your lawn—probably in the middle of the night, so your neighbors don't know you've been cheating—but no one else does so. That way, you get a nice green lawn *and* there is still enough water in the community reservoir. Presumably, the water demands of any one individual—that is, you—are small enough that your private defection doesn't seriously deplete the overall level of the reservoir. The next highest payoff (R, the Reward for mutual cooperation) comes when everyone refrains from watering; admittedly, you get a crispy yellow lawn, but there is also enough water to go around for other, more essential purposes. Then comes a lower payoff (P, the Punishment of mutual defection) if you are a sneaky, cheaty waterer, but so is everyone else; in this case, you may have a lush, green lawn, but only for a short time, after which everyone must deal with water faucets that no longer function. Worst of all is the Sucker's payoff (S), if you are the only one who refrains from lawn watering; that way, you get a crummy lawn plus a community-wide water shortage. As with the standard Prisoner's Dilemma, T > R > P > S.

The key point is that when it comes to social dilemmas, the basic paradox of Prisoner's Dilemma still holds: Whether the others defect or cooperate, each individual's "best" (that is, most rational) move is to defect, but if everyone does this, you are all in trouble. Or, at least, worse off. Economists speak of the "free rider problem," by which individuals are sorely tempted to get something for nothing; when there are lots of free riders, everyone may have to walk.

Sometimes the outcome surprises, and even—in its own way—amuses. During the height of the "terror" in the former Soviet Union, applause for dictator Joseph Stalin's speeches would often last for ten minutes or longer, because no one wanted to be the first to stop clapping! To stop would be to cooperate with the others in the audience, thereby giving everyone a break, but also to defect relative to Stalin himself, such defection being unsafe in the extreme. So, the party faithful found themselves forced to defect—from each other—and keep clapping:

		The audience as a whole	
		stops clapping	keeps clapping
You	stop clapping	R, R	S, T
	keep clapping	T, S	P, P

Only a suicidal Sucker would take the risk of being the person who stopped clapping while everyone else was enthusiastically demonstrating his or her approbation. As a result, Stalin's audience was stuck with the mutual Punishment of seemingly endless applause, although everyone—including, in all likelihood, Stalin himself—would have been better off if all concerned could have relaxed and received the shared Reward of a bit more restraint. (In this case, the "Temptation" to keep clapping was motivated less by hope of gaining advantage over one's fellow clappers than by fear of seeming insufficiently enthusiastic. The Punishment—P—of excessive clapping could presumably be measured in terms of damage to the collective Soviet eardrum as well as sore, reddened palms.)

Nor are such dilemmas unique to the communist world.

In 2000, author Stephen King announced that he would make his next book available online. The *New York Times* reported that "Mr. King is trusting his readers to send him a dollar after each download. . . . If he does not receive payments for at least 75 percent of the downloads, he says, he will stop writing after two chapters, and readers won't learn the end." As it happened, Mr. King's electronic novel, *The Plant,* generated more than 120,000 downloads when the first chapter appeared; by the mid-November installment, only 40,000 were being downloaded, and, of these, only 46 percent were paid for. As a result, Stephen King stopped writing it. "If you pay, the story rolls. If you don't, the story folds," King had written on his Web site. It folded.

Social Dilemmas and the Social Contract

It's a big deal, this business of social dilemmas. After all, individuals only rarely exist in isolation. Nearly always, we interact with the rest of society, expected to cooperate yet tempted to cheat, relying on the cooperation of others yet vulnerable to their defection. The basic concept of "society" assumes give and take, often called a "social contract," whereby individuals make what is essentially a deal with society at large: Each will forgo certain selfish, personal opportunities in exchange for profiting from the cooperation of others.

Lions do it, too. They are notably cooperative, at least within the pride. It often helps to have more than one hunter to kill a large and potentially dangerous animal, such as a water buffalo. There is also a shared benefit if several lions are available to defend the kill from hyenas. At the same time, some lions are less likely than others to hold up their end of the social contract; such laggards are often tolerated, however, because they provide an overall benefit to the rest of the group. Having lots of lions—even if some of them spend most of their time just lyin' around—means that your pride is less likely to be pushed around by other prides, or by packs of hungry hyenas.

Or take the phenomenon known to biologists as "reproductive skew," whereby the dominant members of certain bird flocks permit subordinates to be part of the group—and even, on occasion, to reproduce—so long as they don't get too carried away with it.[3] The alpha birds profit by having the lesser-ranked individuals around, essentially as cannon fodder when it comes to defending the nest, bringing in food, and so forth, while the subordinates find it in their interest to subordinate themselves so long as they can glean some crumbs from the alphas' table. Among birds, this may also include subordinates being allowed to lay a few eggs . . . but just a few.

At the same time, social dilemmas constantly threaten to undermine even the most prosocial of contracts, whether among animals or people. Subordinate birds regularly attempt to lay more eggs than the dominants would like; lazy and even cowardly lions are more frequent than the Wizard of Oz might imagine; and, cynical as it might sound, the likely reason we are told that it is more blessed to give than to receive is that most people, left to their own inclinations, would rather get than give.

Theories of social contract in relationship to social dilemmas have occupied many of the great thinkers in the Western tradition of social philosophy. (Actually, Eastern, too, considering Confucius and his emphasis on social propriety.) The question, in short, is simply this: How to reconcile personal selfishness with public benefit? One of the earliest discussions of this problem can be found in Plato's *Republic*. Toward the beginning of Book II, Glaucon has been claiming that "self interest is the motive that all men naturally follow if they are not forcibly restrained by law and made to respect each other's claims." Accordingly, laws are necessary:

> This is the origin and nature of justice. *[Note, by the way, this is Glaucon speaking, not Socrates, who argues differently.]* It is a compromise between what is most desirable, to do wrong and avoid punishment, and what is most undesirable, to suffer wrong without redress. . . . For anyone who had the power to do wrong and called himself a man would never make such an agreement with anyone—he would be mad if he did.[4]

Thomas Hobbes picked up the argument and lobbied forcefully for the necessity of regulating selfish, nasty human impulses for the good of the larger whole. Although social dilemmas had not been identified as such during Hobbes's time, he clearly saw that the seductive power of social defection was so strong as to be irresistible. In his book *Leviathan*, Hobbes wrote that "during the time men live without a common Power to keep them all in awe, they are in that condition which is called Warre; and such a warre, as if of every man, against every man." The role of the political sovereign, in game theory terms, was to punish noncooperators and keep everyone in line, preventing each from defecting. Hobbes envisioned that without such control, we would all inhabit a state of nature in which people were incapable of arranging for such cooperative endeavors as agriculture, industry, arts, or even society itself, "and which is worst of all, continual feare, and danger of violent death; And the life of man, solitary, poore, nasty, brutish and short."

For Hobbes it was both naive and dangerous to think that one can maximize well-being by simply behaving selfishly, without regard to the situation of others. Thus, he wrote that

> The Foole hath said in his heart, there is no such thing as justice; and sometime also with his tongue; seriously alleging, that every man's con-servation and contentment, being committed to his own care, there could be no reason why every man might not do what he thought con-duced thereunto.[5]

By contrast, the Hobbesian wise man recognizes, in his heart and head, that people must not be allowed to "do what comes naturally," or be encouraged that "if it feels good, do it." This is because defecting comes all too naturally, and what feels good is likely to be bad for the rest of society, and, ultimately, bad also for the most self-indulgent individuals.

Then there is Jean-Jacques Rousseau, who, with his idealization of the "noble savage," seems to be so different from Hobbes. Yet Rousseau, too, in his best-known work, *The Social Contract*, pointed in a similar direction. Rousseau made an important distinction between the "will of all" (the sum of individual desires) and the "general will" (the good of society, taken as a whole), pointing out that the social contract is a way of making sure that pursuit of the former doesn't destroy the latter. Rousseau begins his famous book noting that "Man is born free, and is everywhere in chains." Although he often inveighs against these chains, even Rousseau recognizes that they are necessary:

> In order that the social contract should not be a vain formula, it tacitly includes an undertaking, which alone can give force to the others, that whoever refuses to obey the general will shall be constrained by the whole body: this means nothing other than that one forces him to be free. . . . The undertakings that bind us to the social body are obligatory only because they are mutual, and their nature is such that in fulfilling them we cannot work for others without working also for ourselves.[6]

In short, the peculiar genius of society is that it forces people to abide by their social contracts, and allows them to bypass the siren call of social dilemma, by precisely the kind of restraints and restrictions that Hobbes recommended and that we might expect Rousseau—given his adulation of the "noble savage," untrammeled by rules and regulations, conventions and compulsions—to have opposed. But even Rousseau, apostle of natural human inclinations, recognized the need to say *no* to the Temptations posed by our many social dilemmas.

The same difficulty was identified by the great German rationalist philosopher Immanuel Kant, who argued as follows:

> The problem of organizing a state, however hard it may seem, can be solved even for a race of devils, if only they are intelligent. The problem is: given a multitude of rational beings requiring universal laws for their preservation, but each of whom is secretly inclined to exempt himself from them, to establish a constitution in such a way that, although their private intentions conflict, they check each other, with the result that their public conduct is the same as if they had no such intentions.[7]

Kant recognized that personal defection is socially defective. So he, too, wrestled with the problem of how to protect society—and, thus, each member of society—from the selfish inclinations of its constituents. Kant relied, not surprisingly, on an appeal to rationality ("The problem . . . can be solved even for a race of devils, *if only they are intelligent*"). But there is a problem within the problem, one that even the great Kant, for all his rationality, couldn't solve. It is this: In a world of social dilemmas, the pursuit of rational self-interest results in a bad payoff for everyone.

An early and ingenious experiment investigated how these crummy payoffs arise when individuals literally get in each other's way. The study involved creating something analogous to an escape panic, by presenting subjects with the simple task of removing small aluminum cones from a narrow-necked bottle. Each cone was attached to a string, by which the cones could be extricated; those who removed all their cones were rewarded with cash. The problem was that if everyone pulled his strings at the same time, the exit got clogged and none of the cones could be removed, so no one received any money. In some cases, a time crunch was added: the bottle was slowly filled with water and participants were given extra money if they could remove their cones before they got wet. The rising water level thus represented a fire in a theater, and the cones stood for people in the audience. To rush for the exit (to defect) was equivalent to pulling your string regardless of what others were doing; to cooperate and wait your turn was equivalent to filing out in an orderly manner, whether done by theater patrons or aluminum cones.[8]

Traffic jams were inevitable in this now-classic bottle-and-cone experiment, so long as the subjects had no opportunity to discuss the problem in advance. When the water and financial incentives were removed, however, serious blockages were generally avoided; in this case, the game switched from being a social dilemma to a simple coordination problem, like the Interrupted Telephone Call Game we encountered in chapter 2. In the bottle-and-cone experiment, simple "rules of the road," often improvised on the spot and without any detailed discussion, permitted all the cones to be removed without anyone obstructing anyone else.

At its most fundamental level, civilization itself may depend on people extricating their own cones, without jamming their neighbors'.

The worst Prisoner's Dilemmas are of our own making. If defecting is a Hobbesian tactic, then agreeing to subject oneself to Leviathan's authority or just to the primacy of simple rules may be enough to prevent the nightmare of mutual defection. We agree to forgo our right to loot and pillage (that is, to defect) in return for the Reward of living among cooperators.

In view of its implications, it is not surprising that social dilemmas have generated not only enthusiastic analysis, but also equally energetic criticism. It could be argued that in a social context, Prisoner's Dilemma justifies noncooperation and supports the belief that people will inevitably pursue their own selfish interest. For human society, this view can lead to two responses: designing systems, such as free-market capitalism, that allow people to pursue their self-serving inclinations, since they are going to do so anyway; or mandating restraints and enforcement mechanisms to prevent selfish, socially irresponsible behavior.

It is unlikely that animals have a social conscience. They strive to make the most of their opportunities, feeling neither guilt nor exultation at the outcome. The question of how people perceive social dilemmas is doubtless a function of individual conscience, ethical indoctrination, as well as the likelihood and costs of being caught. (Recall the midnight lawn waterer.) At the same time, responsiveness to such dilemmas may well underlie much of the distinction between liberal and conservative. At the risk of oversimplifying, liberals are eager to cooperate— to their conservative critics, too eager—and thus, willing to subject themselves and others to the risk that some people may cheat, so long as society as a whole might benefit. For conservatives, on the other hand, personal autonomy is crucially important, and we shouldn't be forced to cooperate if we'd rather not. Government has no business, in this perspective, deciding what to do with one's tax money; individuals should be free to make that decision, and to "defect" if that suits them. Conservatives can't stand the notion of getting even the smallest piece of the Sucker's payoff that comes from somebody, somewhere, being a welfare cheat, or another country violating an international treaty.

It is sobering to consider that there may not be a clearly correct approach. Reasonable people can disagree, just as they do on questions of how best to play Prisoner's Dilemma.

However you slice it, personal benefit often clashes with public interest. It is difficult to resist the Temptation to vote for lower taxes, to keep one's own money for one's own use, regardless of the devastating effect such decisions can have on the public good. A game theory perspective helps show that when this happens, the individual may do well pursuing his or her selfish payoff, but usually not for long (once others find out) and certainly not if everyone does the same thing.

All of this runs counter to one of the basic precepts of free-market capitalism, perhaps *the* basic precept. Thus, in the most famous paragraph in his masterpiece *The Wealth of Nations,* Adam Smith introduced the notion of the "invisible hand," whereby the common good is achieved most efficiently when each individual follows private greed:

> It is not from the benevolence of the butcher, the brewer, or the baker, that we expect our dinner, but from their regard to their own self-interest. We address ourselves, not to their humanity but to their self-love, and never talk to them of our own necessities but of their advantages. . . . It is his own advantage, indeed, and not that of society which he has in view. But the study of his own advantage naturally, or rather necessarily leads him to prefer that employment which is most advantageous to the society. . . . He generally, indeed, neither intends to promote the public interest, nor knows how much he is promoting it. . . . He intends only his own gain, and he is in this, as in many other cases, led by an invisible hand to promote an end which was no part of his intention. . . . By pursuing his own interest he frequently promotes that of the society more effectually than when he really intends to promote it.[9]

Political conservatives love this sort of stuff, which gives them permission to extol personal selfishness with the claim that by pursuing private gain, individuals are also promoting the public good. Game theory in general and social dilemmas in particular help point out that this is self-serving nonsense.

It behooves any ardent free-market fundamentalist to give serious thought to this matter. The best—perhaps the only—solution is for external constraints to force individuals to cooperate, because otherwise they won't. But ironically, such constraints go counter to the most basic tenet of unfettered free-market capitalism! I am not going to suggest a way out because as far as I am concerned, the solution is apparent, and at hand: the urgings, requirements, and restraints of society as a whole, that is, the benevolent intervention of government, demanding at least a modicum of cooperation on behalf of society and the greater good. After all, we accept any number of impositions on our personal freedom: It may be in my selfish interest to rob a bank, if I could get away with it. But it isn't in society's interest for there to be lots of bank robbers, so we agree that police, courts, and jails are necessary, to change the payoff matrix and make it disadvantageous for individuals to defect in this particular social dilemma. As we have seen, Hobbes—generally considered a political conservative, incidentally— saw this situation as requiring that individuals hobble their freedom in their own, collective, long-term interest. Doesn't the same argument apply to the worst excesses of unfettered capitalism?

If free-market fundamentalists disagree, it's up to them to point out alternatives.

The Free-Rider Problem

No one likes a free rider. Especially if you've just paid full fare.

The problem, of course, is a social dilemma. Imagine people sneaking under the turnstile for a free ride on a commuter train, occupied by paying passengers. Although there may be some pleasure to be derived from paying your way, even if (and, maybe, especially if) you could have snuck in, on, over, or through without being caught, most people are also inclined to get away with something if they can. Face it: Despite the pleasure of basking in the warm glow of one's own virtue—and, conversely, despite the genuine guilt in violating norms that we have likely been taught—it is tempting to avoid paying, and it is precisely that Temptation that leads most people to resent it when they resist the temptation, only to see someone else sneak aboard without paying.

Economists used to call it the Lighthouse Problem: Assume there are dangerous rocks near the coastline, putting ships, commerce, and lives at risk. Everyone would be better off if there were also a lighthouse, but lighthouses are expensive, and no one person can afford to build one. So who pays for the lighthouse, something that benefits society as a whole but whose cost no one wants to bear? (Or similarly, you might call it the Education Problem, the Hospital Problem, the Firefighter Problem, the Library Problem, the Wildlands Preservation Problem, the National Defense Problem, the Pothole Repair Problem, and so forth.) The answer is that since everyone benefits, everyone should pay, but since no one wants to, everyone must be induced, somehow, to do so.

Karl Marx did not foresee the free-rider problem, clarified as it now is by our recognition of social dilemmas. "They pretend to pay me," went a common wisecrack in the former Soviet Union, "so I pretend to work." Marx also didn't anticipate that efforts to control such defectors would result in a large state apparatus inclined to make use of its coercive powers to govern in its own interest. Although many in the noncommunist West were fond of criticizing Marx for failing to realize that workers would be tempted to cease working if given the opportunity, the fact is that free riders are as much a dilemma for free marketeers as for communist apparatchiks.

Often, reluctance to contribute isn't based on a churlish disregard for others, or on simple, cheapskate selfishness. Rather, the free-rider problem stems from an assumption that turns out to be all too true, all too often: that others won't contribute and therefore, by being the only one to do so, you'll be taken for a Sucker. A remarkable number of people don't mind paying "their fair share," once they are convinced that (1) it is in fact fair, and (2) others are doing the same.

In Jean de La Fontaine's seventeenth-century fable of the grasshopper and the ant, we are told how a thrifty, hardworking ant resents his lazy, free-rider counterpart, the grasshopper, who sings through the summer and then asks for a handout when food is scarce. In a somewhat more upbeat account, *Frederick,* a lovely children's book by Leo Lionni, describes how the eponymous hero, a mouse, sits by dreamily, oblivious to the complaints of his colleagues who are laboring to stock up food for the winter. But eventually, during the cold, dark days that follow, Frederick repays his social debt by warming his fellow rodents with poetic descriptions of the summer sun and flower-strewn meadows. To my knowledge, however, biologists have yet to document cases in which real animals punish freeloaders, à la La Fontaine's ant, or in which genuine Fredericks repay a social debt, whether the currency is seeds or stories. There is evidence, to be sure, that chimpanzees hold interpersonal grudges, and that they punish individuals who have received assistance in the past but refrain from reciprocating when payback time arrives.[10] But this counts more as a failure of one-to-one reciprocity than enforcement of chimpanzee social obligations, by society as a whole.

People, of course, are very serious about checking up on one another's contributions, and blaming those who shirk. Moreover, it is also possible to be too pessimistic, too worried about potential freeloaders; in short, to know too much. (Or to think you do.) Case in point: an experiment carried out among participants at the Third International Conference on Social Dilemmas, held in Groningen, Holland.[11] The forty-three participants—all experts in game theory and social dilemmas—were made the following offer. Each would receive 10 Dutch guilders if the group as a whole would contribute 250 guilders or more. It would be a positive outcome for them all if each would contribute as little as 6 guilders. This would generate 258 guilders in total (exceeding the threshold for the general payout), after which everyone would get 10 guilders, for a profit of 4 per person. What actually happened? They raised 245.59 guilders, and so the potential payoff of 10 per person was lost, along with the money actually contributed.

As noted, the subjects of this informal "study" were attending a conference on social dilemmas and, thus, were experts. It would be interesting to see if nonexperts respond as rationally—and self-destructively— to the identical opportunity/dilemma. (The key may well reside in whether one sees such events as opportunities or dilemmas.)

The Groningen participants were also polled as to their expectation of how much money would actually be raised. Significantly, those who predicted that the threshold amount would be reached contributed an average of 7.24 apiece, whereas those pessimists who predicted that the number would fall short contributed an average of 1.83. Those who cooperate expect cooperation from others; those who don't, don't.* Of course, it may be equally significant that the pessimists contributed anything at all!

One of the important findings that emerged from Robert Axelrod's computer tournament was the importance of being nestled in a network of cooperators. Not that everyone has to be "nice" for niceness to evolve. But there must be at least a few cooperators, in order for the nice guys to benefit from their own niceness. Otherwise, cooperators are Suckers, who eventually disappear. Axelrod's work, and many other studies that he helped to stimulate, demonstrated that cooperation could develop so long as would-be cooperators have at least the opportunity to interact with others who are similarly inclined. In such cases, the benefit of mutual cooperation can be enough to make up for the costs of interacting with the occasional defector, so long as there aren't too many. Another way of saying this: Cooperators are likely to be highly sensitive to the prospect that others are not cooperatively inclined . . . in short, that they are freeloaders or free riders.

There have been many tests of the free-rider problem. For example, a group of twelve subjects may be given $10 each and told that if seven or more contribute to a general fund, all twelve subjects will be given a $20 bonus, whether or not they contributed. If you were one of these experimental subjects, would you contribute? One consistent finding is that cooperation is less likely in large groups than in small ones. Why is this? There are several theories. One has been called the "bad apple" idea. The larger the group, the more likely it is that there will be some noncooperating bad apples, who inspire others to behave similarly. It is hard to refrain from watering your lawn when your neighbors are doing so, or to stay on the sidewalk when all those around you are crossing against the light. Or maybe it's a matter of anonymity in crowds: people feel that their own misbehavior will be less noticed if it is buried in a larger group. Or expectations of reciprocity may simply be lower in a larger group, since TIT-FOR-TAT requires an interaction

*At the same time, it could be predicted that those who were *very* optimistic in their expectation of others' behavior would also refrain from cooperating, figuring that enough money would be contributed—by others—so that they could hold back, and profit additionally by not contributing their share. There are wheels within wheels in this game theory business.

with an identified individual. Or perhaps the key is the sense of personal efficacy, which is likely greater in smaller groups: When there are a hundred million voters, your vote doesn't make much difference. But what about when there are only ten?

The Tragedy of the Commons

A model—first described in a now-classic scientific article by ecologist Garrett Hardin—helps us understand one of the major factors underlying environmental problems.[12] It is also another way of describing a kind of social dilemma. Hardin examined the situation that existed for generations in Britain, in which some grassland was privately owned, while another part, the "commons," was shared property of the community at large. Various citizens owned livestock, such as sheep, which they could graze on their own private land or on the public commons. Even hundreds of years ago, it was well known that overgrazing was harmful to the productivity of the grassland, and so shepherds generally avoided overgrazing their own property. But they treated the commons differently. Landowners recognized that a healthy commons was beneficial to everyone, but each also reasoned that if he refrained from grazing his sheep on the commons, then others would doubtless take advantage of this restraint, and fatten their own flock on the public lands. As a result, tendencies to be prudent and ecologically minded were suppressed because individual sheep owners reasoned that if they did not take advantage of the commons, then surely someone else would. So, if the commons was going to be degraded in any event, they may as well be the ones who do it. The result was deterioration of the commons, until it was no longer fit to support sheep, or shepherds.

Here is the situation:

		Everyone else	
		saves the commons	uses the commons
An individual shepherd	saves the commons	R, R	S, T
	uses the commons	T, S	P, P

The highest payoff goes to the individual shepherd who uses the commons while everyone else is saving it. He is tempted (T) to defect, hoping to fatten his flock at the public trough, while everyone else—by refraining—assures that the commons itself remains in reasonably good shape; at the same time, the defecting shepherd is responding to his fear of getting the lowest, Sucker's payoff (S) if he had helped preserve the commons by keeping his sheep off it, only to have everyone else use it. The second highest payoff, the Reward for mutual cooperation (R), comes if our chosen shepherd elects to preserve the commons and everyone else does, too. This way, the commons is saved, to everyone's long-term benefit. The next-to-worst payoff is the Punishment of mutual defection (P), if the individual shepherd defects and so does everyone else.

The tragedy of the commons, then, is that each individual, seeking to gain personal benefit, finds himself engaging in behavior that is to the *dis*advantage of everyone. It can also be generalized to other resource-related difficulties, whenever short-term selfish benefit conflicts with long-term public good. For example, there may be short-term, self-centered benefit to a factory owner in using the atmosphere as a public sewer; after all, even if his effluents pollute the air, this cost is borne more or less equally by everyone who breathes, whereas the factory owner personally is saved the expense of having to install pollution-control devices.

We have already looked at the social dilemmas involved in overuse of scarce resources. It may be inconvenient to recycle and, in fact, easier for individuals simply to throw their garbage away, or to use more than their share of scarce commodities, such as water, electricity, and so forth. After all, each of us may derive some personal gain or enhanced convenience by overconsuming while the cost—in overcrowded dump sites or worldwide resource shortages—is by contrast a diffuse and general one, borne by all. Besides, each consumer is likely to argue that if she doesn't abuse the environment, then surely someone else will (which is just what the sheep owners told themselves about the commons).

The tragedy of the commons has global dimensions: Scandinavian forests and lakes suffer from acid rain because of the outpourings from

English smokestacks, while Britain gets the economic benefit of burning its coal. Psychologist Andrew Colman points out that this particular example isn't properly a social dilemma, because the countries of Scandinavia aren't able to make choices in this matter; they are simply the victims of English avarice. "A better example," notes Professor Colman,

> might be over-fishing in the North Sea. Herring, which used to be plentiful all round the British coast in the 19th and early 20th centuries, were fished to extinction. Now cod are becoming endangered for the same reason. It's in every fisherman's individual interest to catch as many cod as possible, but if they all defect in this way, then the cod (and the fishermen!) will soon go extinct like the herring before them.[13]

Regrettably, there are other social dilemmas of the herring/cod sort: Japan and Norway consistently defy international outcry while hunting the world's great whales to the verge of extinction . . . after which Japanese and Norwegian whalers will follow their quarry into oblivion. Brazil seeks to benefit economically from the Amazon rain forest, even though such "benefit" requires that it be destroyed, to the ultimate detriment of everyone, including Brazilian loggers.[14]

What to do?

Most people agree that the tragedy of the commons is a genuine tragedy, not in the Aristotelian sense of something that we experience vicariously and that provides a kind of emotional cleansing, but a dangerous and even horrifying reality. Observers derive different take-home messages from it, however, depending on their political persuasion. Thus, conservatives see the tragedy of the commons as confirming the need for private property rights. If each shepherd had his own plot of land, they argue, it would be treated better. Good stewardship, under this view, comes from having a personal stake in the resource. Any resource, including land, that is owned in common might as well be owned by no one, since it will be cared for by no one.

At the same time, liberals point to the same tragedy as demonstrating the need for restraint, and since such restraint is unlikely to come from the shepherds themselves, it must be supplied by laws, backed up by social (that is, governmental) power. In this clash between individualistic benefit and collective rationality, Garrett Hardin argued for the latter, enforced by "mutual coercion mutually agreed upon." (Shades of Hobbes's *Leviathan* and Rousseau's *Social Contract*.)

During the period of communist-style governments in the Soviet Union and eastern Europe, the environment was no better protected there than in the capitalist West; indeed, by most measures, it was worse. Socialist production goals were accorded the highest priority and, as a result, environmental values suffered greatly. International meetings of environmentalists were notable for the hopes of the participants, often pinned unrealistically on the opposing system. Thus, Western environmentalists expected that state control and authority would generate models of more reliable environmental protection, while environmentalists from the Soviet bloc had an equally idealistic—and unrequited—hope that private ownership might lead the way toward rational ecological stewardship.

It has been said that under capitalism, people exploit people, whereas under communism, it's the other way around! Either way, the environment gets abused.

Rousseau's Stag Hunt

I've tried to emphasize that game theory helps clarify ancient problems, most of which still exist today. In some cases it may suggest solutions, although as we have seen, the preferred take-home message often depends on one's preexisting bias. Thus, as we have seen, those inclined to a free-market ideology will likely interpret the tragedy of the commons as a call for private ownership, whereas those preferring collectivist solutions are prone to see the same tragedy as requiring social—that is, governmental—intervention.

One of the more useful conceptualizations of this sort comes from French social philosopher Jean-Jacques Rousseau. It doesn't suggest an answer, but it certainly helps us visualize the problem.

Rousseau asked his readers to imagine a small group of hunters pursuing a stag. Such a large animal offers plenty of meat for the hunters, whose success, in turn, requires the cooperative efforts of several participants. (It might help to think of pretechnologic, Pleistocene hunters, using bow and arrow, instead of modern hunters using high-powered rifles. It will likely take many arrows to bring down the prey, so success depends on everyone playing his part.) Rousseau added this complication: While engaged in the hunt, one of the hunters sees a rabbit. Here is a Temptation. Admittedly, a rabbit is small compared to a stag, even compared to a stag divided among, say, four or five hunters. Pursuing the rabbit therefore offers a smaller payoff than does sticking with the stag. But the rabbit is, to switch metaphors, a bird in the hand, whereas the larger hunting party might not even encounter its preferred prey. And even if their quarry is eventually located, other hunters might have been similarly tempted to go off after their own rabbits, leaving too few left to bring down the stag. As a result, a hunter who spots a bunny would be sorely tempted to take off after it, leaving the others shorthanded.

Here are Rousseau's own words, from *A Discourse on Inequality*:

> If it was a matter of hunting deer, everyone well realizes that he must remain at his post; but if a rabbit happened to pass within reach of one of them, we cannot doubt that he would have gone off in pursuit of it without scruple and, having caught his own prey, he would have cared very little about having caused his companions to lose theirs.

And here is a 2 × 2 game theory matrix for the Stag Hunt, showing both alphabetic payoffs (T, S, R, and P), to emphasize its defect/cooperate component, and numeric payoffs (4, 3, 2, and 1) to show their relative magnitudes:

		The others	
		hunt the stag	chase a rabbit
One hunter	hunts the stag	4, 4 (R, R)	1, 3 (S, T)
	chases a rabbit (T, S)	3, 1 (P, P)	2, 2

For Rousseau's individual hunter, the highest payoff comes from sticking with the original plan, and hunting the stag . . . provided others do the same (and provided, of course, that a stag is encountered and killed). Unlike a Prisoner's Dilemma, in a Stag Hunt the cooperative outcome is highest: R, the Reward of mutual cooperation when everyone does his "social duty" and hunts the stag. If you think it is sufficiently likely that the others will stick it out and hunt the stag, you'll do so, too. The Stag Hunt is not a Prisoner's Dilemma. (Remember that in Prisoner's Dilemma, and social dilemmas as well, the highest payoff comes from defecting no matter what the others do; in Stag Hunt, the highest payoff comes from cooperating so long as the others do the same.)

Nor is Stag Hunt precisely a social dilemma. But it is nonetheless a dilemma, since, as with Prisoner's and social dilemmas, there is also a temptation to defect—to chase a rabbit—if you fear that others will also be defecting. After all, even though the payoff of bagging a rabbit (3) is less than that of being part of a successful stag hunt (4), it is nonetheless better than the lowly return garnered by any Suckers who continue forlornly hunting the stag while the rest of the party has peeled off chasing rabbits (1).* As with Prisoner's Dilemmas, the consequence of everyone chasing rabbits (P, or the Punishment of mutual defection), is less desirable—a mere 2—than the outcome if everyone had cooperated and stuck with the stag, and with one another. In the Stag Hunt, therefore, $R > T > P > S$.

As with other two-person games of this sort, the Temptation to defect is influenced by many things beyond pure logic. Or we might say that "pure logic" in such cases involves factoring in many considerations beyond the simple payoff matrix. In particular, each player should be concerned not only with the theoretical payoff as depicted in each quadrant of the matrix, but also with the probability that this payoff will actually be generated. For example, you might well be tempted to chase a rabbit if you see a lot of them or if they seem especially vulnerable (perhaps because there has been a recent snowfall, making them easier to track), since your chance of success is likely to be higher. In addition, increased availability and vulnerability of rabbits would also increase your Temptation because it raises the

---

*I've also assumed that the payoff of bunny hunting when others continue stag stalking (3), is higher than when others also go after a bunny (2), simply because too many bunny hunters might well lower the success rate of our initial defector; but it really doesn't matter.

likelihood that your fellow hunters have been similarly tempted. And if they are more liable to give in, you are better off following the same course.

You would also be more prone to rabbit chasing if you doubt the character or the stag-killing skill of your fellow hunters, or even if you have reason to worry that they are less rational than you, in which case it would be paradoxically rational of you to preempt them by behaving irrationally first! In Kipling's poem *If*, the poet sings the praises of "keeping your head while all those around you are losing theirs." A laudable goal, except in something like a Stag Hunt, in which keeping your head and pursuing the stag while all those around you are getting theirs turned by images of plump, easily caught bunnies, turns out to be a losing policy.

At the same time, there are considerations that predispose toward sticking with the stag: anything, for example, that makes R bigger or S smaller. There may be social opprobrium associated with being known as an inveterate bunny chaser. Other hunters may not invite you along on their next excursion. And you'd probably not be welcome at any celebratory venison dinners. Nonetheless, the Stag Hunt is a powerful phenomenon. Testimony to its impact can be found in myriad cases of animal behavior, in which wolves, for example, often lag behind when it comes to making that first—and most dangerous—attempt to grab a cornered moose, or when penguins mill about at the edge of an ice floe, each waiting for someone else to jump in first and thereby reveal whether a hungry leopard seal is lurking just beneath the water.

You may have noticed that we've already encountered a one-on-one variant of the Stag Hunt: the reassurance game involving India and Pakistan in chapter 3, the only difference being that the Stag Hunt is a social dilemma. Interestingly, the Stag Hunt dilemma can be ameliorated by reassurances of various sorts. Hunters tell themselves how manly and reliable they are, sharing stories of past exploits and making much of their reputations. They sing the praises of stag dinners (both kinds!), and speak glowingly of the hunt's likely success.

Here is another variant on the Stag Hunt, first suggested by Douglas Hofstadter.[15] Imagine that there are many people—say, a dozen or so—each sitting in a small isolation booth, each equipped with a button. If everyone sits quietly for ten minutes *without* pressing his or her button, everyone gets $1,000. Sounds easy, but add this complication: If anyone presses his or hers, that person will get $100, whereas everyone else gets nothing. It still sounds easy. Just do nothing and everyone still gets $1,000, which beats the $100 earned by the would-be defector, and certainly is better than $0. Cooperating, then, is still the smart thing to do. Smarter yet, however, is to realize that somebody just might be so stupid or so distrustful as to push a button, in which case you would end up with nothing; so maybe you had better push yours first. And the smartest people of all, we might think, realize that if the others are neither stupid nor distrustful, some of them are likely to be very smart—just like you—and, hence, they are reasoning similarly, which makes it even more imperative that you push *your* button before they push theirs.

If so, then the outcome of all this infernal rationality is financial disaster for everyone, just as the misbehavior of Rousseau's imaginary hunters spells meatless meals for the Suckers who cooperate and stick with the stag.

Author William Poundstone suggests that a traditional naval mutiny might also be a Stag Hunt in disguise. The whole crew would be better off if the mutineers succeeded in getting rid of the odious Captain Bligh (that is, if everyone hunts Bligh, the human equivalent of a stag). But if too many defect (that is, if too many remain loyal to the captain—which is to say, opt to chase the rabbit) then the mutineers will be in big trouble. What often happens in real-life attempted mutinies is that individuals hesitate and dither, seeking to get a sense of what others are going to do before committing themselves. (Recall the tentative behavior of stickleback fish, discussed in chapter 3, who assessed each other's willingness to confront a dangerous predator.) Political and military coups often have this flavor, as does surrendering during wartime.

There has been a great deal of research on decision making by individual animals, very little on groups. At least one study, however, concluded on the basis of mathematical models that "democratically" made decisions should lead to better overall outcomes than those arrived at "despotically." Not that the animal equivalent of voting need be called for; rather, like our purported mutineers, individuals should simply watch and see what the others are doing.[16]

Another example: Everyone is better off wearing motorcycle and bicycle helmets, but until wearing them was legally required, people were reluctant to do so, feeling that it marked them as sissies. Make it law that everyone hunt the stag, or that no one push his or her button, or that everyone must wear a helmet, and most people comply. In addition, whether they admit it or not, they have good reason to be grateful as well.

A Bit of Psychology: Groups and Grouches

You may have noticed something conspicuously missing in most of our discussion of social dilemmas (and, for that matter, Prisoner's Dilemma) so far: psychology. In a sense, behavior is being predicted and evaluated as if the participants aren't really people at all, but rather, computers, or automatons whose actions are motivated by a clear, cold-eyed calculus of payoff-maximization, and nothing else.

We'll look more at this in chapter 7. For now, it's worth pointing out that most of game theory was developed by mathematicians, logicians, and economists, who, in fact, are especially predisposed to see people as logical, predictable, self-benefiting, calculating machines. (By the way, biologists make the same assumptions about animal behavior, and, as we'll see in chapter 6, they are generally proved correct.) When it comes to human behavior, on the other hand, the expectation of perfect rationality makes for good—that is, internally consistent and, thus, satisfying—theory and gratifying mind games, but may leave something to be desired when it comes to how real folks, out in the messy world, actually live their lives.[17]

Social dilemmas were first identified in a book titled *The Logic of Collective Action,* published in 1965 by economist Mancur Olson, which, in turn, had been stimulated by Hardin's discussion of the tragedy of the commons.[18] Notably, both Olson and Hardin were pessimistic about the potential of people to dig themselves out of social dilemmas. According to Hardin,

Each man is locked into a system that compels him to increase his herd without limit—in a world that is limited. Ruin is the destination toward which all men rush, each pursuing his own best interest in a society that believes in the freedom of the commons. Freedom in a commons brings ruin to all.

Such pessimism isn't only a product of twentieth-century scholarship. "What is common to the greatest number gets the least amount of care," wrote Aristotle, in his *Politics*. "Men pay most attention to what is their own; they care less for what is [held in] common."

When psychologists have gotten into the fray—something that has only happened seriously in recent years—they have usefully muddied the waters, adding a bit of human complexity to Olson and Hardin's logical constructs and the dire warnings of political philosophers from Aristotle to Rousseau. Fortunately, for those of us who worry about the dangerous, seductive lure of social dilemmas, the pictures that psychologists paint have been considerably more hopeful, even as they add some maddening complications.

First of all, even though social psychologists recognize the formal similarities between social dilemmas and Prisoner's Dilemma, they are also inclined to point out some important differences. In Prisoner's Dilemma, for example, the cost of defection is borne entirely by the other player; in social dilemmas, it is distributed across a large group, including the player herself, which could make the social opprobrium of defection somewhat greater. Also, because social dilemmas are played, by definition, against a social group, the decision to cooperate or defect is anonymous, or nearly so—assuming you don't get caught. By contrast, each participant in a Prisoner's Dilemma is unambiguously connected to his or her most recent move. On the other hand, in repeated games of Prisoner's Dilemma, each player has at least the possibility of influencing the behavior of the other, not directly of course (the nature of "games" requires that players have no immediate control over each other), but indirectly, by rewarding—either positively or negatively—the other player's behavior. In repeated games of Prisoner's Dilemma, for example, a cooperator can be rewarded in the next game by receiving cooperation, just as a defector can be punished by a retaliatory defection. Such a relationship, fundamental to strategies like TIT-FOR-TAT, is difficult to obtain in social dilemmas.

According to psychologists who study the phenomenon, anonymity is important. Several decades ago, a young woman named Kitty Genovese was brutally attacked and murdered by a knife-wielding assailant. Despite her screams, which were heard by several dozen people in her New York City neighborhood, no one intervened. Not even to call the police. This failure of "bystander intervention" generated a wealth of research by social psychologists, eager to understand why people didn't help. In a sense, each bystander was faced with a social dilemma: whether to intervene or not, and even, at much less personal cost, whether to phone the police or not, while not knowing what others might do. One finding: The larger the bystander group, the less likely that an individual will "cooperate." Everyone thinks—or hopes—that someone else will take the lead, thereby conveying a whiff of coordination games, such as the Interrupted Telephone Call, although with more sinister consequences. With large bystander groups, responsibility can be so diffused as to be essentially absent altogether. (It usually isn't terribly serious, by contrast, if no one initiates a callback after an interrupted phone conversation.)

Closely related is what psychologists call "inconsequentiality," the idea—often self-serving—that each individual has only a negligible effect on the overall outcome. Not surprisingly, appeals for charitable giving typically emphasize the importance of the individual, just as public service announcements encouraging people to vote point to the fact that "every vote counts." (The 2000 U.S. presidential election, especially in Florida, may have given special impetus to this argument.) In our personal lives, most of us experience—at least to some extent— the opposite of inconsequentiality; namely, personal efficacy. We may have no direct impact on how the United States conducts its economic or foreign policy, or what is done to fight AIDS, but we can determine what clothing we wear, what we eat at a given meal, whether we watch television or read a book at night, whether or not to raise a finger.

Numerous studies have shown that people are willing to cooperate in a social dilemma (give something up, take somewhat less, volunteer their time) in direct proportion as they believe that their actions will actually have an effect. This is a genuine challenge, since the reality is that in most cases, personal efficacy really does decline as group size increases. (Remember Sartre's dictum?)

Or recall Andrew Colman's bad apple theory: The larger the group, the more likely that it contains at least a few bad apples. Even more devastating, people may well *expect* that large groups are especially likely to include such defectors, so it may as well be them. Add to this the observed phenomenon that people are exquisitely sensitive to indications that others are cheating. For example, in a study designed to simulate resource depletion (based on the tragedy of the commons, sometimes called a "take-some" game), it was found that people's assessment of one another's behavior was crucially important in influencing whether they would refrain from taking more than their fair share. If subjects believed that a common resource pool was being depleted because of natural, environmental factors, they were more likely to cooperate and take less themselves. But if they believed that depletion was caused by the greed of other players, they were less likely to cooperate, and more likely to defect and overconsume, just as they felt the other group members were doing.[19] (We have already considered the alternative—"give-some" games—such as the guilder-giving experiment at the Groningen social dilemma conference.)

Finally, there is the matter of "social loafing." It is one thing to loaf around on a Sunday afternoon, doing nothing all by yourself, and quite another to fail to pull one's weight in a social situation, when part of a supposed group effort. In a well-known study of social loafing, subjects were asked to shout "Rah!" with all possible force, several times during the experiment. This may seem silly, but, in fact, it requires considerable effort and even mild discomfort to do so, and indeed, seeming silly may be yet another reason why most subjects were reluctant to bellow as requested. Participants wore headsets and blindfolds, so they could neither see nor hear whether they were alone or part of a group. The energy expended in their yells was assessed by a decibel meter, and the results were striking: "When individuals thought they were combining their efforts with one other person they shouted at only 82% of their alone performance, and this dropped to 74% when they thought they were performing with five others."[20] Other researchers have consistently confirmed the general pattern; people are prone to social loafing.

It isn't clear whether animals are similarly tempted, but it seems likely that they are. What is clear is that animals in social groups often do less than when they are solitary. Ostriches, for example, don't stick their heads in the sand; in fact, they spend quite a lot of time looking around, scanning the horizon for predators. When in a flock, each ostrich spends significantly less time looking around than when he or she is alone.[21] Is this social loafing? Or is it one of many legitimate reasons for being social in the first place?

In any event, it seems unavoidable that, for human or animal, cooperation is, if anything, harder to come by in social dilemmas—when individuals are playing against a group—than in one-on-one Prisoner's Dilemmas. But not so fast. For one thing, there is the matter of group identification, or, as it is sometimes called, "we-ness." Individuals, for all their individualism as well as their inclinations to goof off and avoid being suckered by other goof-offs, also have a habit of identifying themselves as part of a group. And the stronger the identification, the less likely one is to defect. This helps explain why countries work so hard to instill a sense of patriotism, of national identity, a willingness to subordinate personal goals and needs to the benefit of the larger group. In fact, people seem downright eager to make such associations, whether at the level of root, root, rooting for the home team or giving one's life as a patriotic martyr. Even a suicide bomber can be seen, albeit counterintuitively from the perspective of the victim, as someone who "cooperates" with his group, willing to accept the Sucker's payoff of certain death, even though presumably he would be better off if someone else volunteered his life for the cause. But, for the self-proclaimed martyr, the payoff of self-destruction is, in a sense, very high, insofar as by doing so he guarantees himself a place in heaven. (Recall Pascal's Wager.)

People are social creatures, influenced by things other than a simple, rigid, payoff-maximizing calculation. They also have big brains, which come up with "reasons" for doing—and not doing—things that aren't always amenable to reason. As Pascal noted, "the heart has its reasons that reason does not know." Economists and mathematicians typically take an undersocialized and overrationalized perspective on human behavior, since people are always embedded in social relationships of some sort, typically spiced with a hefty dose of emotion as well. In short, psychology counts, for better or worse.

To some degree, it is also a matter of *personal* psychology, the idiosyncratic inclinations of individuals, presumably due in part to their genetic tendencies as well as their own experiences as they grow up. Maybe some people are just "naturally" selfish and inclined to defect, while others are equally predisposed to be generous and cooperative. But in either case, experience also counts. It has been said, only partly in jest, that a conservative is a liberal who has been mugged. Similarly, maybe a defector is a cooperator who has had one too many encounters with other defectors. Some psychological research suggests that, in effect, there are two types of people: those inclined to cooperate (who eventually learn that others vary in their degree of cooperativeness), and those who habitually defect. These defectors attribute noncooperativeness to everyone else but are unlikely to learn their error because they consistently defect, which evokes defection from others, many of whom would have cooperated had they encountered someone less nasty, cynical, paranoid, or, in the case of social loafing, lazy.

Also, don't overlook the role of social punishment, administered by other "players" who stand to lose if someone in their group is stingy or a shirker. As we've already discussed, one response to observing such a cheater is to be more likely to cheat oneself—recall how difficult it is to wait at a crosswalk when all those around you are jaywalking. But what about the alternative: making the cheater pay? This raises the problem of who administers the rebuke, retaliation, or Punishment. Ideally, it should be everyone, but this simply opens the opportunity for yet more defection, on the part of individuals who stand back and let others do the punishing! One answer is for Punishment to be meted out not only toward defectors, but toward anyone who refrains from punishing them. Next step, then, is to punish those who won't punish those who defect, and so on, ad infinitum. In the close-formation battle phalanxes favored by the Roman legions, every foot soldier depended on the man alongside him, to provide protection via his shield. Desertion in battle was a capital offense, to be administered on the spot; moreover, anyone who failed to kill a deserter was himself subject to immediate death!

In most practical cases, however—when the Punishment for defection is considerably less drastic—people have been found to mete out a kind of "altruistic Punishment" quite readily. Thus, in one study, experimental subjects played a game in which the group as a whole did better in proportion as each individual player was generous, but players also had the opportunity to hold back on their personal contribution and take advantage of the payments made by others. After each round, the players were informed what the others had just done. When provided the opportunity to punish freeloaders, such that by giving up one unit of their own payment the freeloader would be docked three units, 84 percent of the experimental subjects did so at least once. That is, they willingly took money out of their own pockets in order to punish shirkers. More than one third did so five times or more, and just under 10 percent exceeded ten times.[22]

These results can be seen as revealing something nasty and spiteful about people: they will go out of their way—even enduring personal financial loss—just to be mean to someone else. But in addition to this "glass-half-empty" interpretation, there is also a "glass-half-full" counterpart. Thus, the fact that people will punish a cheat in a public-goods situation, even if doing so may be costly, is, in a sense, evidence for a kind of altruism. By maintaining social norms at their own cost, "punishers" are being essentially unpaid policeman, willing to incur costs while making a kind of citizen's arrest. Interestingly, the results of this self-appointed policing are beneficial for the group as a whole: The researchers found that when such Punishment was not allowed, cooperation within the group quickly broke down, but when it was permitted, individuals cooperated more as the rounds of play continued.

In short, people are readily inclined to turn on cheaters (anyone who has ever bridled at the boor who breaks into line at a ticket window or supermarket cashier can easily understand) and there is much to be said for it, not least because social pressures—including the possibility of direct retaliation—make such cheating less likely.

Individual psychology offers some additional reason for optimism. Think about a symphony orchestra. More specifically, imagine that you are one of eighteen musicians playing in the second-violin section. What difference does it make if you play off key, or, indeed, if you don't play at all? Admittedly, it doesn't take much additional effort to do your best once you have already shown up for a performance, but in theory there is little reason for you to exert yourself. And yet, no one has ever experienced an orchestra performance in which the musicians are all silent, each expecting to be a free rider on the efforts of everyone else! In truth, serious orchestral musicians often wonder if they, individually, can be heard, but at the same time, something keeps them from defecting.

It is hard to say precisely what this might be, but it doubtless has to do with the fact that individuals often take great pride in contributing to something larger than themselves, something worthwhile and, if possible, beautiful. Musicians—to continue the example—achieve their position in an orchestra via competitive auditions. Moreover, serious performers are often highly competitive in themselves, and they can be fiercely protective of their reputations, both individually and collectively. Musicians are also typically committed to their art, such that it would be unthinkable for them to sabotage a performance out of social loafing or personal selfishness. What is true of orchestra musicians may well be true, differing in detail but similar in pattern, for the rest of us. When it comes to the merits of "illogical" cooperation, few people are entirely tone deaf.

A recent study suggests that there may even be some underlying brain mechanisms at work.[23] The brain activity of thirty-six women was monitored using functional MRI while they played games of Prisoner's Dilemma. The basic finding was that mutual cooperation led to activation of brain regions—the nucleus acumbens, the caudate nucleus, ventromedial frontal/orbitofrontal cortex, and rostral anterior cingulate cortex—that are associated with the pleasurable sensations of "Reward." Cooperation feels good. In addition, as part of the research, sometimes the subjects were told that they were playing against a computer; at other times, against other women. The subjects cooperated more in the latter cases, and when they did so, their brains' pleasure centers were more active, leading to the conclusion that there may be specific brain mechanisms that reward cooperation *with other people.* All this makes sense. After all, our species has been exposed to a deep dilemma: If we cooperate, we can often reap long-term Rewards, but to do so, we must frequently overcome the Temptation to defect in the near term. So maybe our brains have been wired to take the long view.

In short, maybe there is something in the human brain (or, at least, in the brains of women: this experiment has not yet been repeated with men) that provides us with the neural equivalent of a pat on the head— rather, *in* the head—encouraging us to "do the right thing."

Some More Psychology: Lifeboats, Loners, and Losers

Here is a generalization: When it comes to social dilemmas, public choices tend to be more cooperative than private ones. Anonymous donors are rare. (Character, it is said, is what you do when no one is watching!) Furthermore, the greater the salience of one's group identity, the greater the inclination to behave cooperatively. Not surprisingly, therefore, the more isolated a person feels, the more likely he is to defect.

Beyond the question of group identification, there is also the matter of group norms, notably ethical and moral expectations. Part of the socialization process is being told what sort of behavior is expected of us; in fact, this may be most of what socialization is about. Moreover, maybe the danger of defection, so aptly captured in the models of social dilemmas, goes a long way toward explaining why socialization happens at all. In this view, socialization equals indoctrination, the preferred doctrine being that individuals should submerge their selfish inclinations for the benefit of the group.

The German philosopher-genius-madman-poet Friedrich Nietzsche was probably the most articulate spokesperson for the cynical view that society encourages self-sacrifice because the unselfish Sucker is an asset to others:

> Virtues (such as industriousness, obedience, chastity, piety, justness) are
> mostly *injurious* to their possessors. . . . If you possess a virtue . . . you

are its *victim!* But that is precisely why your neighbor praises your virtue. Praise of the selfless, sacrificing, virtuous . . . is in any event not a product of the spirit of selflessness! One's "neighbor" praises selflessness because *he derives advantage from it.*[24] [italics in original]

If Nietzsche is correct, then there is probably a distressingly manipulative quality to morals, to most religious teachings, to the newspaper headlines that celebrate the hero who leaps into a raging river to rescue a drowning child, to local Good Citizenship Awards and PTA prizes.

"That man is good who does good to others," wrote the seventeenth-century French moralist Jean de la Bruyère. Nothing objectionable so far; indeed, it makes sense (especially for the "others"). But de la Bruyère goes on, revealing a wicked Nietzschean cynicism:

If he suffers on account of the good he does, he is very good; if he suffers at the hands of those to whom he has done good, then his goodness is so great that it could be enhanced only by greater suffering; and if he should die at their hands, his virtue can go no further; it is heroic, it is perfect.

Such "perfect" heroism can be wished only on one's worst enemies.

Another interesting psychological finding is that people are more likely to cooperate when they trust that other people will also cooperate.[25] No one wants to be the sole cooperator, not because misery loves company, but rather because being a lonely Sucker—given the nature of social dilemmas—means that you are, in fact, liable to be truly miserable (remember that the Sucker's payoff is substantially lower than the Reward of mutual cooperation . . . and the latter is, in a sense, the outcome when everyone is a Sucker, which means that no one is). So, not surprisingly, part of the socialization process involves instilling confidence in the reliability that others will cooperate along with you, combined with exhortations concerning what is right or wrong in your own behavior.

People are widely urged to be kind, moral, altruistic, and generally cooperative, which suggests that left to their own devices, they would be *less* kind, moral, altruistic, and generally cooperative than is desired.

At the same time, it is also common to give at least lip service to the precept that people are fundamentally good. "Each of us will be well advised, on some suitable occasion," wrote Sigmund Freud, in *Civilization and Its Discontents*, "to make a low bow to the deeply moral nature of mankind; it will help us to be generally popular and much will be forgiven us for it." Why are people generally so unkind to those who criticize the human species as nasty and selfish? Maybe because of worry that such critics might be seeking to justify their own unpleasantness by pointing to a general unpleasantness on the part of others. And maybe also because most people like to think of themselves as benevolent and altruistic or, at least, to think that other people think of them this way. A cynic is harder to bamboozle.

In *Civilization and Its Discontents,* perhaps his most pessimistic book, Freud went on to lament that one of education's sins is that

> it does not prepare them [children] for the aggressiveness of which they are destined to become the objects. In sending the young into life with such a false psychological orientation, education is behaving as though one were to equip people starting on a Polar expedition with summer clothing and maps of the Italian Lakes. In this it becomes evident that a certain misuse is being made of ethical demands. The strictness of those demands would not do so much harm if education were to say: "This is how men ought to be, in order to be happy and to make others happy; but you have to reckon on their not being like that." Instead of this the young are made to believe that everyone else fulfills those ethical demands—that is, that everyone else is virtuous. It is on this that the demand is based that the young, too, shall become virtuous.

On the other hand, the expectation that others will be aggressive, nasty, and prone to defect can become a self-fulfilling prophecy, especially if it leads people to be aggressive, nasty, and prone to defection in anticipation.

Here is an example of a laboratory study of social dilemmas conducted by psychologists. It has been called a Public-Goods Game, since it investigates the following situation: Individuals have the opportunity to relinquish something of value—think of paying taxes—with the understanding that if enough people do so, everyone will benefit, although those who choose not to relinquish will benefit even more, so long as others cooperate. (The Groningen experiment, described earlier, in which game theoreticians were the subject, was a classic public-goods dilemma, also known as a "give-some" game.) In one such study, variants of which are often conducted in social psychology labs, nine experimental subjects are each offered a check for $5, which they can cash immediately (that is, they can defect and gain $5). Or, they can cooperate by relinquishing their $5. The rules state that if a certain minimum number of people (say, a total of five) agree to give up their original stake, everyone gets a bonus of $10, while noncontributors end up with $15 (the $5 they decided to keep, plus the $10 gained because the minimum total was collected). The cooperators thus end up with a total of $10, assuming that there are enough of them, as compared to zero if only a few individuals cooperate. At the same time, defectors can receive $5 if they cash in their money right away, or $15 if they cash in *and* a sufficient number of *other* people choose to cooperate.[26]

It's an interesting concept. Here is the payoff matrix:

		Number of cooperators	
		0 to 4	5 to 9
Individual's choice	cooperate	$0	$10
	defect	$5	$15

Notice that for each individual, the payoff for defecting is higher than for cooperating, regardless of how many others choose to cooperate. If zero to four cooperate, defectors get their $5, whereas cooperators get nothing; if five to nine cooperate, then defectors get $15, whereas cooperators get only $10. Clearly, it's a kind of Prisoner's Dilemma, in which the rational thing is for each individual to defect— to take the $5—but the overall effect (the shared Punishment of mutual defection) is hurtful for everyone.

Yet, when a group of social psychologists ran this experiment, about 60 percent of participants chose to cooperate, as a result of which, about one-half the experimental groups received the $10 per person bonus.[27] So the prospect of cooperation is far from hopeless. Particularly noteworthy was this additional finding: When the same experiment was run under a "full communication condition"—that is, when the participants were given an open-ended opportunity for group discussion prior to each individual making his or her private choice—a whopping 100 percent of the groups obtained the bonus! That is, the problem of defection entirely disappeared.

How was this achieved? In most cases, the groups agreed to some sort of lottery procedure, by which five individuals were randomly chosen to cooperate. By doing so, they admittedly received a net return of "only" $10 (the bonus alone, since cooperating meant that they gave up their initial $5), while the others got $15 (the bonus plus $5). Either regime was better, however, than the $5 that defectors would have gotten or the zero that cooperators would have been stuck with if there weren't enough of them. Lest this seem too obvious, bear in mind that each decision—to cooperate or defect—was made in privacy, so that no one could be confronted subsequently with the onus of defecting.

Even more interesting, perhaps: When asked later to explain why they chose as they did, most participants in the communicating, cooperating lottery said something like "because that was what the group decided" rather than "because it was in my interest to do so." And this was true even though the group had no way of knowing who subsequently did what, or to enforce its "decision." Remember, the final "moves" by each individual were made privately, so every participant was free to be an anonymous defector.

Several decades ago, social psychologist Stanley Milgram conducted a renowned study of "obedience to authority," in which he demonstrated that Americans were so willing to follow orders that they would deliver what they thought were painful and even life-threatening electric shocks to innocent subjects. Milgram's research generated much soul-searching about the dangers of following orders. But hidden within that cloud of susceptibility may be a silver lining, if it means that psychology inclines us to follow orders for the good of the group—and ultimately, ourselves—once that group makes such a decision.

Next, let's take a leap to a similar if more gripping social dilemma, the Lifeboat Game. As described by psychologist Sanford Braver, it is based on this kind of situation:

Nine survivors of a shipwreck are adrift at sea in a lifeboat. The boat contains enough provisions that, with careful rationing, could keep all the survivors alive for ten days. An island, known to contain additional provisions and communication equipment, is known to be fifteen days away. The survivors know that rescue is impossible until and unless the island is reached. One of those aboard proposes that the survivors draw straws with three short and six long straws. If each of those drawing the short straws jumps overboard to certain and immediate death, the lowered consumption will enable the remaining six to stretch the provisions for the full fifteen days and be rescued.[28]

This Lifeboat Game gives rise to the following payoff matrix:

		Total number jumping	
		0 to 2	3 to 9
Individual's choice	jump	immediate death	immediate death
	don't jump	10 to 14 days starving on lifeboat, then death	rescue

In this gruesome "game," the rational strategy is to agree to the proposed lethal lottery (since otherwise you are sure to die), then to refuse to jump if you draw a short straw, all the while insisting, if you drew a long one, that anyone with a short straw must do as previously agreed! It isn't clear whether situations of this sort have ever actually occurred, and if so, whether short-straw people actually honored their commitments, "because that was what the group decided." After all, leaping voluntarily from a lifeboat isn't quite the same as forgoing $5.

At least one real-world situation came horribly close to this simulation. The nineteenth-century survivors of the whaleship *Essex,* sunk by a sperm whale (and which ultimately provided the model for Herman

Melville's *Moby Dick*), faced a similar dilemma. As recounted in a gripping book by maritime historian Nathaniel Philbrick, the lifeboat-bound survivors, facing certain starvation, drew straws to decide who was to be cannibalized by the others.[29] We can imagine that it was somewhat easier to agree to this "rational" solution in principle than to sacrifice oneself—that is, to cooperate—if push came to shove. It may therefore be noteworthy that the cooperative outcome in this case wasn't left in the hands of the unlucky individual. Instead of being asked to jump overboard, he was shot . . . although as survivors recounted it, the victim did, in fact, cooperate rather than protest or struggle.

Psychologist Braver, a professor at Arizona State University, carried out a laboratory simulation that captured at least some of the logic of the Lifeboat Game, albeit without its maritime drama or life-and-death epic (and before Philbrick's book was published). Here, everyone was given $5, which they could keep or return. If three or more chose the latter (cooperative) path, the keepers (defecting) would get $15 each, but in any case, the cooperators got nothing. Compare this matrix with the Lifeboat Game:

		Total number returning their $5 stake	
		0 to 2	3 to 9
Individual's choice	return the $5	$0	$0
	keep the $5	$5	$15

In this experiment, the subjects were told that their final decision would be made privately, and that they would be spared any subsequent interactions with the other players. Thus, no congratulations for cooperating, no recriminations for defecting. Each group was also given the opportunity to discuss their strategies freely, during which considerable attention was devoted to "promises" and "swearing" to follow the group decision. In every case, this came down to holding an impartial lottery, and then agreeing to abide by its outcome. The bottom line: Although narrowly defined "self-interest" directs that everyone drawing a short straw would defect and keep the $5, more than 70 percent freely returned the money to the experimenter. Adherence to the group—even a group that had no meaningful existence and disbanded immediately with the conclusion of the experiment—trumped the financial incentive.

Quite likely, a larger incentive (say, $500 instead of $5) would have generated more defection. By the same token, greater persistence of the group (say, a reunion planned for a week later) might well have induced more cooperation. It seems clear in any event that social obligations are an important part of any social dilemma, and that—humanely manipulated—they may also suggest a way out.

Finally, let us end on a note both horrifying and ennobling. When terrorists hijacked four civilian airliners on September 11, 2001, eventually crashing two of them into the World Trade Center and one into the Pentagon, they were evidently foiled in their plans for airliner number four. It crashed into a rural part of Pennsylvania, apparently after spirited resistance by some of the passengers, who attempted to retake control. We will probably never know precisely what happened during that awful time, but it is clear that through use of cell phones, the passengers became aware of the successful attacks on the World Trade Center that had just occurred, and they knew that their own hijackers had similar intentions for themselves as well. From further cell phone conversations, it is also clear that several of the passengers, unarmed, decided to rush the hijackers in a desperate bid to save themselves and—if at all possible—to prevent yet another monumental catastrophe.

It is a true-life re-creation of the lifeboat situation described above. Assume that the passengers knew that if they did nothing, they would surely die. Seen this way, fighting back is only "logical," since it entails choosing a high probability of death over a certainty. In fact, it doesn't seem to merit any fancy analysis at all, just posthumous congratulations for doing the right thing, not only for themselves but also for other innocent would-be victims, on the ground, who would otherwise have died had the hijackers been allowed to succeed.

But stay with this horrible example a bit longer, because it does, in fact, have a painful and instructive game theory component. Thus, there is still a "defector" strategy available to the passengers: hold back, or get someone else to do the fighting rather than run the immediate risk yourself. After all, the terrorists were armed with razor-sharp blades; the resistors had only their bodies and their courage. There must have been great Temptation to be a freeloader, to let others take the chance of being butchered by the hijackers, since if the resistors ended up being successful, any nonresistors would be saved, too, but without running the risk of being fatally slashed in the process. Yet it appears that at least four passengers rose to the challenge, behaved cooperatively—and valiantly—and died (along with everyone else on board) in the attempt.

It is quite possible to interpret such extraordinary examples of courage and self-sacrifice as responses to a social dilemma, whose awful dimensions were quickly perceived and then acted upon. Thus, the four passengers who evidently acted so nobly were all young, male, and relatively athletic or physically imposing. They may have also been imbued with a sense of patriotism, self-sacrifice, personal efficacy, and/or social responsibility, which effectively would have changed the payoff matrix under which they acted.

Or maybe this is "over the top," taking rational analysis too far, or into realms where it simply doesn't apply, giving off an unpleasant reek of cynicism rather than the much-ballyhooed clear air of cognition. Maybe it is when dilemmas are greatest that we act most strongly from our hearts rather than our heads. And maybe that's precisely what made the actions of those people on board United Airlines flight 93 so extraordinary, and so worth our gratitude and admiration.

5

Games of Chicken

"We're eyeball to eyeball," said Secretary of State Dean Rusk, "and the other fellow just blinked."

It was 1962, and Rusk was describing the Cuban Missile Crisis, when the world came closer to nuclear war than ever before. It was also a near-perfect example of a Game of Chicken. In that terribly tense time, leaders of the United States (and, presumably, the Soviet Union) were painfully aware of their need to anticipate what the other side was likely to do. "We discussed what the Soviet reaction would be to any possible move by the United States," said JFK's adviser, Theodore Sorenson. "What our reaction would have to be to that Soviet reaction, and so on, trying to follow each of those roads to their ultimate conclusion."[1]

Fortunately, that "ultimate conclusion" did not include the annihilation of human life. Game theory does not deserve to be congratulated for this positive outcome. But it does deserve to be taken seriously, if only because it was relevant then, and still is.

The Classic Case

The Game of Chicken is the last of the four major 2 × 2 games (we've already looked at the other three: Battle of the Sexes, Leader, and Prisoner's Dilemma). Like the others, the Game of Chicken is defined by its payoffs in a 2×2 matrix, and we'll get to that. But first, let's look at the classic example of Chicken, a situation so readily grasped, and so horrifying, that in some ways it transcends the need for logical analysis.

Two cars are speeding toward each other. Each driver can do one of two things: swerve or go straight. To win, you must go straight; the one who swerves is the "chicken." Of course, it is possible for both drivers to swerve, in which case both are "chickens," but neither suffers in reputation relative to the other. Or, and here is the crunch, literally: Both drivers can go straight, each attempting to win and neither willing to lose. In this case, both lose out, possibly forever.

It is said that Games of Chicken were first played by California teenagers during the 1950s, although this may simply be an urban legend. It is clear, however, that philosopher Bertrand Russell saw the gruesome parallel between such games and nuclear brinkmanship: Each side wants the other to back down—to turn aside—although neither is willing to do so itself. A head-on collision beckons.

A modified Game of Chicken was memorably portrayed in the 1955 movie *Rebel Without a Cause*. The archetypal rebel/hero, played by James Dean, competed with a rival. The winner's payoff? A beautiful girl. In this case, the two competitors drove their cars toward a cliff, instead of toward each other. The last one to bail out of his vehicle wins. (Dean's rival got his sleeve caught and couldn't escape, so he "won," which is to say, he lost.) Although the *Rebel Without a Cause* version is logically a Game of Chicken, too, it is easier, more dramatic—and in some ways, more true to life—to think in terms of two cars bearing down upon each other, each wanting the other to swerve. The payoff matrix for the Game of Chicken is shown below:

		Driver 2	
		swerves	goes straight
Driver 1	swerves	R, R	S, T
	goes straight	T, S	P, P

In such a game, to go straight is to defect and to swerve is to cooperate. The driver who defects—that is, who perseveres in going straight ahead—wins. He gets T, the Temptation to defect, but if and only if the other cooperates; that is, if he swerves, and thus gets S, the Sucker's payoff of being a chicken. If both drivers swerve, each gets R, the Reward of mutual cooperation; no one wins, but no one loses, either. The dramatic outcome in a Game of Chicken, however, and the one that draws our eyes and our imaginations, is P, the Punishment of mutual defection if both drivers go straight ahead. It is a punishing outcome indeed.

The Game of Chicken looks a bit like Prisoner's Dilemma, since it poses a similar conundrum: If both players try to get the best possible payoff for themselves (T), they end up with an outcome that is punishing for them both (P). In both the Game of Chicken and Prisoner's Dilemma, the highest payoff, T, comes when a player defects and the other one cooperates (so that the cooperator gets S). And in both the Game of Chicken and Prisoner's Dilemma, the next highest payoff, R, comes when both players cooperate. But here is the difference: In Prisoner's Dilemma, the Punishment of mutual defection (P), although less than R, is not all that bad; it is better than S, the Sucker's payoff. But in the Game of Chicken, the Sucker's payoff is better than P. Much better. Better to swerve and let the other guy get the girl than for both of you to go straight, and, in so doing, go straight to the hospital, or to your death.

In mathematical symbols, the relationship of the payoffs in Prisoner's Dilemma is $T > R > P > S$. In the Game of Chicken, it is $T > R > S > P$. (To complete the picture for the four most interesting 2×2 games, the payoffs for Leader are in this relationship: $T > S > R > P$, and for Battle of the Sexes, it is $T > S > P > R$.)

The Game of Chicken is the most dramatic situation analyzed by game theory, although it is less intellectually stimulating than Prisoner's Dilemma.* (At least, it hasn't given rise to as much research attention.) Nonetheless, Games of Chicken are important, even when they aren't as life-threatening as in the case of love-crazed, testosterone-poisoned teenagers. Or ethically challenged adults with their fingers on a nuclear button.

---

*Drama is in the eye—or the brain stem—of the beholder. Thus, Prisoner's Dilemma can be plenty dramatic, as is the Ultimatum Game, or the Dollar Auction. Most people would agree, however, that Chicken is the most *dangerous* game.

So far, I have concentrated on aspects of game theory that have particular resonance in life's events and have gone easy on mathematics, proofs, and theorems such as minimax (you can't get much easier than omitting them altogether!). But in describing the Game of Chicken, it will help to take a detour into this territory . . . and besides, this book is based on the proposition that interesting ideas are fun. Most important are Nash equilibria and Pareto points, the former named for John Nash, a brilliant but mentally ill mathematician who has undergone a remarkable recovery and was awarded the Nobel Prize for economics in 1994, and the latter referring to the work of turn-of-the-century Italian economist and sociologist Vilfredo Pareto. Although sometimes associated with the early development of fascism because of his theory that socioeconomic elites are naturally superior, Pareto also gave his name to a notably egalitarian concept. Thus, he established the foundation of modern welfare economics, suggesting that a society's resources are not optimally distributed so long as it is possible to make even one individual better off without making anyone else worse off.

A Nash equilibrium in a two-player game, such as Chicken, is a pair of strategies, each of which is the best response to the other; each strategy gives the player using it the highest possible payoff given the other's strategy. Often, in analyzing games, we don't know for certain what the other player will do; in some cases, however, even after you know the other's move, you still have no reason to change your own. Such a case is a Nash equilibrium. It has this particular virtue: It is stable. Think of an uppercase V as if it were a physical structure. A Nash equilibrium represents the position of a tiny ball within that V: any departure from that lowest point would be unstable, so our hypothetical ball, even if jiggled, would end up where it started. Prisoner's Dilemma has a Nash equilibrium; namely, mutual defection. If you know that the other player is going to defect, your best move is to defect. If you know that the other player is going to cooperate, your best move, still, is to defect. Not ideal, in the world of social behavior, but stable. Your best strategy is clear, regardless of what the other player does or is going to do.

Whereas Prisoner's Dilemma has a single Nash equilibrium (defect-defect), Chicken has two: defect when the other player cooperates, and cooperate when the other player defects. Think about it. If you're play-ing Chicken and you know the other player is going to cooperate, you do best by defecting. If you know he will defect, you do best by cooper-ating. Knowing the other player's move, there is no doubt what you ought to do.

Pareto is different. Here, both players—together—are doing the best that they can, given the rules of the game. Compared to a Nash equilibrium, the Pareto version (sometimes called Pareto optimality), is much more precarious. This time, imagine an inverted V, with the ball very tentatively balanced at the apex: ∧. At a Nash equilibrium, no one is inclined to move from his or her existing position, whereas Pareto offers a more nerve-racking kind of balance, since if either player devi-ates from his or her current strategy, the other is typically worse off.

Actually, there are some Pareto equilibria in which one player's devi-ation can improve the lot of the other, and others in which deviation by either player makes everyone worse off. As a result, Pareto-type games can sometimes be very stable and, thus, very simple, so they don't receive much attention. Take, for example, the following payoff matrix:

		Individual A	
		cooperates	defects
Individual B	cooperates	100, 100	0, 0
	defects	0, 0	0, 0

Defection by either player would cost the other one dearly. But it would be equally costly for the defector. The payoffs make it unlikely that anyone will defect, therefore creating a situation in which they consider each other partners rather than competitors. However, Pareto points garner more attention when they are unstable (perhaps for the same reason that car crashes make the news, rather than reports that "there have been no accidents today").

Mutual cooperation (R, R) in Prisoner's Dilemma is a Pareto point, a condition in which both players do well, but each is tempted to switch. In Chicken, mutual cooperation (swerve, swerve) is also a Pareto point, in that if you know the other player is swerving, you'd be well served by going straight. Nash equilibria are stable (what you'd expect from equilibria!) because if either player moves, that player finds himself worse off; there is no temptation to move away from a Nash equilibrium. Pareto points, on the other hand, are vulnerable to efforts at self-aggrandizement.

A bit more terminology: In Prisoner's Dilemma, defection is what game theorists call a "dominant" strategy, because it is always better than its alternative—in this case, cooperating—regardless of what the other player does. In Chicken, neither has a dominant strategy, since his or her best move depends on the other. The point is that when playing Chicken, there is no evident best move. (Do you think it might be best to go straight? Not if the other player also goes straight! Perhaps, then, it's better to swerve? Not if the other player also swerves.)

At Nash equilibrium, no one regrets his or her chosen move, *given what the other did.* This latter part is important. In the case of Prisoner's Dilemma, for example, if both cooperate, each will be inclined to defect if given a second game, whereupon both would "achieve"— more accurately, be stuck with—a Nash equilibrium. Having arrived there, both players may be consumed by regret, but what is regretted is that the other side did what she did, not that—given what she did—you did what you did.

Is that clear?

Fortunately Games of Chicken are comparatively easy to figure out, at least when it comes to understanding their dynamics. But surviving them is a different matter. Unfortunately, "swerve, swerve" is not a stable Nash equilibrium. It is merely an unstable Pareto point, which tempts each player to reach for a better outcome: if the other player swerves, you do better going straight. Swerving may seem the rational strategy for any Chicken player, but if you know (or believe) that the other fellow is a swerver, you cannot avoid being tempted to go straight. The rub is that he is thinking the same thing.

Some Examples: Chickens, One to One

Whenever two parties can be described as being on a "collision course," you are likely witnessing a Game of Chicken. It's not clear that genuine barnyard chickens play chicken, but, as we'll see, lots of other animals do. Picture two neighboring but not especially neighborly dogs, growling and barking at each other across a fence, each making itself seem large and fierce, daring the other to come so much as a step closer, threatening to escalate immediately and watching each other like, well, chickens. Similar games lurk behind international crises, for example, or labor-management disputes, situations in which mutual escalation can result in a catastrophic outcome (strikes, for example, or wars). Or face-offs of other sorts. In 1995, the Republican-controlled Congress, emboldened by its electoral success the year before, played a Game of Chicken against President Clinton, threatening to shut down the U.S. government unless they got their way. Clinton made the same threat. A kind of train wreck occurred, but political commentators agree that the damage was not mutual: Newt Gingrich and company came out worse, because the American public blamed them (correctly, it appears) for having caused the collision.

A classic Game of Chicken, recounted by Homer in *The Iliad,* was brought to the attention of game theory buffs by economist and nuclear strategist Thomas Schelling.[2] Menelaius and Antilochos are neck and neck in a chariot race. They are speeding along—in the same direction—when the track ahead suddenly narrows. One or the other or both of them must swerve, or else risk an accident. Whoever swerves has in essence conceded the race to the one who defects, who persists on a straight-ahead course. Here is how Homer tells it:

> Menelaius was driving in the middle of the road, hoping that no one would try to pass too close to his wheel, but Antilochos turned his horses out of the track and followed him a little to one side. This frightened Menelaius, and he shouted at him: "What reckless driving, Antilochos! Hold in your horses. This place is narrow, soon you will have more room to pass. Otherwise, you will foul my car and destroy us both!"

This is Homer's Hold Your Horses Game, with 4 being the highest outcome, and 1 the lowest (you can also substitute T = 4, R = 3, S = 2, and P = 1):

		Menelaius	
		holds his horses	drives ahead
Antilochos	holds his horses	3, 3	2, 4
	drives ahead	4, 2	1, 1

As Homer recounts the tale, when Menelaius shouted at Antilochos, urging him to hold his horses, Antilochos "only plied the whip and drove faster than ever, as if he did not hear," whereupon Menelaius "fell behind: he let the horses go slow himself, for he was afraid that they might all collide in that narrow space." So Antilochos won, but Homer argued that he did so "by trick, not by merit."

Antilochos' trick was to pretend that he couldn't hear Menelaius' shouted warning. We'll see later that this can be a good tactic indeed. So is plain old-fashioned courage, as long as the other player believes in the would-be defector's bravery, although, as Gabriel Laub has pointed out, "When everyone is courageous, there is reason to worry."[3] After all, heroes rely on meeting cowards: What would become of the courageous if they were always matched with others, equally brave (or foolhardy)? Another way to look at it: Heroic perseverance rarely pays off if your opponent is equally determined to be a hero. Something's got to give.

Unfortunately, questions of judgment and heroism don't arise only in the world of Greek heroes and larger-than-life politicians. Consider this one, played between an average burglar and an equally mundane homeowner. Call it the Gun Game:

		Homeowner	
		doesn't keep a gun	keeps a gun
Burglar	enters without a gun	2, 2	1, 4
	enters with a gun	4, 1	3, 3

The Gun Game points out the uncertain line that often separates Prisoner's Dilemma and Games of Chicken. Thus, we might call it the Gun Game à la National Rifle Association (NRA), since it argues that the lowest payoff for the homeowner arises if the burglar has a gun and the victim doesn't, while the highest payoff is to have a gun when the burglar doesn't have one. It also suggests that it is better for both parties to have a gun than for burglar and homeowner to be unarmed. The result is a Prisoner's Dilemma, "resolved" in a logical but—to my mind, at least—socially unacceptable way by everyone carrying a gun. This game encapsulates the perspective of the National Rifle Association because in effect it whitewashes the mutual defection answer to the social dilemma of gun ownership, by claiming that it isn't really a negative outcome at all. "An armed society," according to the NRA, "is a polite society."

Alternatively, there is the Gun Game, à la Brady Bill:

		Homeowner	
		doesn't keep a gun	keeps a gun
Burglar	enters without a gun	3, 3	2, 4
	enters with a gun	4, 2	1, 1

In this case, which, I believe, is more realistic, the worst payoff is if *both* burglar and homeowner have guns (lower right corner). Then, like Menelaius and Antilochos, or President Clinton and the 1995 Republican Congress, or James Dean and his rival, a Game of Chicken ensues. "If guns are outlawed," says the NRA, "then only outlaws will have guns." More likely, if guns remain legal and gun ownership stays unrestricted, nearly everyone will have guns, in which case the chickens are likely to be armed and dangerous.

Worse yet, just as homeowners will have to assume that all burglars are armed, burglars will similarly assume that all homeowners are armed. As a result, those burglars who aren't deterred by the prospect of homeowners bristling with handguns are likely to preemptively act on that assumption, which means that not only would they carry a weapon, but they would also feel pressured to be the first to use it. "Two scorpions in a bottle," was a description applied to the United States and the Soviet Union during the worst days of the nuclear-nervous Cold War. A mutually armed, Chicken-playing society threatens to make this scorpion scenario an everyday domestic reality.

Pushing the game to its natural conclusion, from the question "to pack a pistol or not," we quickly arrive at "to shoot or not." What happens when an armed homeowner detects a burglar (who may or may not be armed)? Should he use his weapon or refrain? And what about the burglar? (Recall the Gunslinger Game from chapter 1.) A similar dilemma can be described for the decision "to resist or not to resist," which must be made by the victim of a mugger, and "use weapon or don't use weapon" for the assailant. Here, it isn't clear which outcome is better—or rather, least bad—from the victim's perspective: to resist a mugger who has a weapon, or to succumb to your assailant's demands. The traditional advice from police authorities is to avoid resistance: When faced with the choice "your money or your life," most people opt to give up their money. (A famous skit featuring the late comedian Jack Benny was a notable exception. Benny, whose public persona included being a skinflint, responded to the mugger's impatiently repeated ultimatum with "I'm thinking! I'm thinking!")

For the rest of us, the Mugger/Victim Game is closer to Chicken than to Prisoner's Dilemma, and, in such cases, discretion (that is, "swerving" or "not resisting") is the better part of valor. This is precisely what most animals do. Those ill-tempered dogs that we left growling and posturing at each other are likely still at it; their barks are usually worse than their bites.

Although a train wreck is especially likely to get our attention, the reality is that people often find themselves involved in Games of Chicken that are less dramatic, or potentially catastrophic, than head-on collisions, wars, ruinous labor strikes, or potentially lethal shootouts. Imagine, for example, that you and your roommate both notice that you are running out of dishwashing detergent or getting low on beer, or that the garbage is piling up and needs to be emptied. Each of you would prefer that the other do the deed (buy some soap, pick up some beer, or take out the trash). And so you may both enter into a kind of slow-motion, low-intensity Game of Chicken, each hoping that the other will swerve. In the worst case, neither does so, and you both lose some quality of life, but, thankfully, not life itself.

Detailed studies of animal behavior have allowed biologists to eavesdrop on the private lives of different species as never before. And yet, the frustrating truth is that we still don't know nearly enough about the subtle, delicate interactions that undoubtedly pervade the lives of our animal cousins. The dramatic events—fights, sex, murder, rape (equivalent in their own way to train wrecks and head-on collisions)—are increasingly acknowledged. Prominent among the things that we don't know, on the other hand, are the innumerable details of how animals work out their low-intensity, game-theoretic issues, equivalent in their own ways to roommates playing their Garbage Game.

Life—human and animal—is more complicated than models, partly because even when the behaviors are evident, it can be unclear what game is being played. Take, for example, marital fidelity: At one level, of course, it is a matter of love, commitment, devotion, and morality. But it is also true that each partner's behavior is likely to influence the other's. Insofar as this occurs, monogamy could be a Game of Chicken, in which the worst payoff is for both husband and wife to defect, endangering the marriage. But this also implies that each player's goal—his or her highest payoff—is reached by defecting (having an affair), while the other cooperates (doesn't). A dark prospect indeed.

It is well known that "marital" infidelity is very common in the animal world. It is not known, however, whether animals use the threat of infidelity as a kind of blackmail, with one or both members of a mated pair keeping the other hostage in this manner. On the other hand, animals have been known to engage in the equivalent of human "retaliatory affairs," as when female red-winged blackbirds leave their harem ground and mate with an outsider male while the "harem keeper" is off patrolling his realm and looking for additional conquests.[4]

One way to win a Game of Chicken is to threaten to defect, hoping to induce cooperation. In unstable marriages, the threat of an affair may have this effect, but seems unlikely to be a powerful predictor of domestic bliss. Similarly, even though the Game of Chicken predicts Nash equilibria in which one player defects when the other cooperates, an adulterous, cheating mate combined with another who remains faithful would not seem to portend great stability, or happiness. At least, not among humans, capable of being both reflective and jealous.

Interestingly, one could view monogamy as an iterated sequence of games, rather than a one-shot affair. (After all, monogamy is supposed to involve a series of interactions between husband and wife, continuing—by definition—over time.) As a series of iterated Games of Chicken, each partner would feel pressure to cooperate (be faithful) in order to maintain the reward of mutual cooperation from his or her partner in the future. Fear that his or her defection would bring about defection would motivate partners to stay honest, avoiding the lowest returns—including a likely divorce—for both. A dispiriting prospect, perhaps, but no less valid for that.

Maybe the best monogamous game, however, is a simpler one, and not a Game of Chicken at all. In a good marriage, both husband and wife get the highest payoff from mutual cooperation, the lowest payoff from mutual defection, and intermediate payoffs from defection by one or the other:

		Wife	
		is faithful	is unfaithful
Husband	is faithful	4, 4	2, 3
	is unfaithful	3, 2	1, 1

Such a relationship would be stable at mutual cooperation (i.e., fidelity), which sounds nice but is belied by the simple fact that defection by one partner, or both, is estimated to apply in more than 50 percent of married couples.[5]

Finally, in case anyone is getting the impression that Games of Chicken are always simple affairs (or always lead to affairs), consider this version, suggested by Anatol Rapoport.[6] In it, Chicken is merged with Battle of the Sexes, and each player has three possible moves, instead of just two. A man and his wife disagree about their vacation: He wants a cruise, she wants mountains, both could tolerate seashore. Neither wants to go anywhere alone. Here is a possible matrix, in which, as usual, 4 is the highest payoff and 1 is the lowest:

		Wife goes to		
		mountains	seashore	cruise
Husband goes to	mountains	2, 4	1, 1	1, 1
	seashore	1, 1	3, 3	1, 1
	cruise	1, 1	1, 1	4, 2

One seeming compromise given these payoffs might appear to be vacationing at the seashore (each gets 3). But what if the husband announces that he has already bought tickets for a cruise, and his wife would be more than welcome to come along? Such a take it or leave it strategy puts the wife in a bind. If she leaves it, she gets a mere 1, whether she goes, solo, to the mountains or the seashore. If she takes it, she gets at least 2. But is she obliged to capitulate? Not necessarily. What if she makes a counter-ultimatum, also a fait accompli: "Do what you want, but *I'm* going to the mountains." Then both may wind up with the lowest payoff, 1. In effect, the Battle of the Sexes has become a Game of Chicken.

Situations of this sort happen often, if not in private life then certainly in politics. The president threatens to veto legislation that isn't to his liking. Congressional leaders threaten to pass it anyhow, hoping, if necessary, to override any veto. Eventually, cooler heads prevail and a compromise is reached.

At other times, the competitive environment of Games of Chicken simply brings out the worst in people. An especially well-known research example is the "trucking game" devised by social psychologists Morton Deutsch and R. M. Krauss, and presented to the scholarly community with the revealing title "The Effect of Threat upon Interpersonal Bargaining."[7] In this simulation, two players represent competing trucking companies, each attempting to deliver its cargo as quickly as possible. Both have the choice between two routes, one of which offers the players a substantially shorter trip. The catch is that the short route includes a stretch of single-lane road, and the two trucking companies are traveling in opposite directions! As a result, if both attempt the same short route at the same time, each blocks the other (defect, defect). Both players lose valuable time until one decides to back up (that is, back down and "cooperate") and take the longer route. This is better than remaining gridlocked, but it gives the other company a "victory."

In certain diabolical variations, one or both players control a gate that can block the other from proceeding beyond a certain point. Without gates, both players quickly learn how to cooperate by alternating use of the shorter route. But when gates are available to both, gridlock is common; moreover, things typically do not get better over time. When only one subject controls a gate, the results are intermediate. One interpretation of all this is that when threats are available, people behave more competitively. If so, this finding has important implications for international affairs, especially deterrence and political, economic, and military intimidation generally.

The trucking game is neither precisely a Prisoner's Dilemma nor a Game of Chicken. It is more complex than it seems. But so is life.

Madmen, Steering Wheels, and Other Ploys

Prisoner's Dilemma is especially frustrating because the rational move (in single games, or in iterated games with a definite end point) is to defect, yet that strategy is less than optimal for both players. But at least there is a "solution" to the dilemma, as irksome as it may be. When it comes to Games of Chicken, by contrast, there is no single best strategy. At least none that is mathematically or even logically valid.

So, let's turn to psychology. Earlier, we looked at some of the factors that induce people to cooperate or defect when playing Prisoner's Dilemma. But what about trying to win? We've already considered TIT-FOR-TAT, which suggests a way of winning, or, at least, of not losing. Among the recommended strategies for people who find themselves stuck in a Prisoner's Dilemma, convincing the other player of one's trustworthiness turns out to be especially prominent. After all, if you want the other player to cooperate, it helps if you can assuage his anxiety that you might defect. And so, people playing Prisoner's Dilemma often send reassuring, confidence-building signals. "Trust me, I'm a reliable fellow." "I've always cooperated in the past." Or maybe, "I understand, as well as you, the benefits to both of us if we cooperate."

A Game of Chicken is different. Here, the greatest fear isn't being suckered, but suffering the Punishment of mutual defection. The most effective message, therefore, is not to reassure the other player that you will cooperate, but, rather, the exact opposite: threats that you really and truly are going to defect.

To see this more clearly, let's compare the archetypal Game of Chicken between individuals (two drivers heading toward each other) with Games of Chicken between countries (the best-known example being the Cuban Missile Crisis). There are some remarkable similarities.

How does a player win a Game of Chicken between two cars? Accept, right off, that there is no way to guarantee victory. The best either player can hope for is to improve his or her odds of coming out ahead, that is, of inducing the other player to swerve. As to tactics, there are many, although none are especially appealing. For one, reputation matters. If you are known as a nonswerver, perhaps because you have been "undefeated" in a number of previous encounters, your opponent is bound to take that into account. Not surprisingly, national leaders have long been concerned that their country have a reputation of standing by its commitments, just as Richard Nixon, during the Vietnam War, was worried that the United States not be seen as a "weak, pitiful giant."

Reputation can be burnished in several ways, one of the most intriguing being this: Cultivate an image of being literally crazy, or, better yet, suicidal. Imagine that you are playing Car Chicken against someone who gets into his vehicle laughing maniacally, swearing that he isn't afraid to die, or even welcomes it. Such a person would seem especially unlikely to swerve; therefore, the most logical response on your part is to do so. Whether one is actually insane or simply faking it, there is a payoff to convincing your opponent that you have taken leave of your senses. The game theoretician might therefore give this advice to someone about to play Car Chicken: Cultivate a reputation for ferocity, even insanity, since whoever is—or seems—craziest, wins.

One of the most terrifying sights in the animal world is an elephant in "musth." Huge bulls, oozing a weird greenish glop from their eyes, behave with crazed, violent abandon, taking risks and defying the basic rules of pachyderm propriety (and also, incidentally, giving rise to the term "rogue elephant"). The effect of musth on others speaks volumes about the reason for this peculiar phenomenon in the first place: facing an elephant in musth, people—not to mention other elephants—are well advised to get out of the way rather than confront a creature that is temporarily crazy. That is the point. A "musthing" elephant is "crazy like a fox," since by signaling that it is irrational and unpredictable, it is likely to get its way.

This, in turn, leads to the intriguing suggestion that human emotions might serve a comparable function. Blushing, yelling, becoming uncontrollably red-faced with rage: these and other signs of temporary insanity may well have evolved because of the strategic benefits they can convey.

On the international level, Daniel Ellsberg referred to the "political uses of madness." And Richard Nixon, when president, actually attempted to reap a benefit by convincing the North Vietnamese that he was so crazy that his irrational threats of the U.S. employing nuclear weapons must be taken seriously. During the Vietnam War, Nixon advised his chief of staff, H. R. ("Bob") Haldeman:

> I call it the Madman Theory, Bob. I want the North Vietnamese to believe that I've reached the point where I might do *anything* to stop the war. We'll just slip the word to them that 'for God's sake, you know Nixon is obsessed about Communism. We can't restrain him when he's angry—and he has his hand on the nuclear button' and Ho Chi Minh himself will be in Paris in two days begging for peace.[8]

It didn't work, but it is nonetheless chilling to note that Nixon had seriously attempted to employ his madman theory. It is also noteworthy that game theory, developed as a means of applying high levels of rationality to reveal optimum tactics and strategy, ends up—in this case at least—recommending behavior that is so irrational as to be literally insane.

There is another variant on the madman theory, based on the same goal: convincing the other player that you are unwilling or literally unable to swerve, leaving the other player no choice but to do so. Here is the renowned nuclear strategist Herman Kahn, in his influential book *On Escalation:*[9]

> The "skillful" player may get into the car quite drunk, throwing whisky bottles out of the window to make it clear to everybody just how drunk he is. He wears very dark glasses so that it is obvious that he cannot see much, if anything. As soon as the car reaches high speed, he takes the steering wheel and throws it out of the window. If his opponent is watching, he has won.

The Game of Chicken becomes a contest to see who can be the first to discard the steering wheel! To be sure, there are problems with such a strategy, at least two of which were acknowledged by Kahn himself: "If his opponent is not watching, he has a problem; likewise if both players try this strategy."

When the invading army of William the Conqueror burned their own ships in 1066, it sent a clear message: We will not retreat—because we cannot!—and so the only option is to fight and win. A similar strategy was followed by Hernando Cortés, who had all but one of his ships destroyed after landing in Mexico. It is reported that during the Trojan War, the defending Trojans attempted to burn the Greeks' ships, which would have been highly counterproductive had they succeeded; in fact, we might have expected the Greeks to have beaten them to it! In short, burning your bridges behind you can be an effective strategy, since, like throwing away your steering wheel, it constitutes an irrevocable commitment, in the face of which a foe might wisely back down.

Nuclear strategists have long been troubled by this problem of "credibility," painfully aware that it tends to be in short supply when the threat is nuclear retaliation after having been attacked, since nothing is to be gained by obliterating an opponent who has already attacked you, and much is to be lost (the livability of the planet). And so, strategists—strongly influenced by game theory, it must be noted—responded by recommending a policy of "launch on warning," whereby nuclear retaliation would be assured by computers, taking the prospect of rational intervention by human beings "out of the loop." In Stanley Kubrick's spectacularly funny, black humor movie *Dr. Strangelove, or how I learned to stop worrying and love the bomb,* the Soviets installed a doomsday device, wired so as to automatically destroy the world in the event of a nuclear attack on the USSR. Their strategic misstep: They neglected to tell the Americans! (Think back to Kahn's warning that the driver who discards his steering wheel "has a problem" if the other driver isn't watching.)

There are several ways of convincing the other driver that you aren't going to swerve. Reputation continues to play a role. Your determination to go straight varies with your desire to be victorious; that is, the more important the victory, the greater the commitment to the strategy and, hence, the more risk taking. In the case of Car Chicken, imagine that James Dean was head-over-heels in love with the girl (the game's "payoff"), while the other player wasn't so—literally—crazy about her. Who is more likely to swerve?

In the realm of international affairs, it is widely assumed that countries are more likely to persevere in a Game of Chicken if their national interests are directly engaged; that is, if they are seriously "in love" with the payoff outcome. During the Cuban Missile Crisis, the strategic interest of the United States was more at stake than was that of the USSR, since Soviet missiles in Cuba would have threatened U.S. security more than their removal would have endangered the Soviets. When the Soviet Union invaded Czechoslovakia in 1968, no one seriously expected the United States to intervene militarily, since events in Eastern Europe were critically important to the Soviets, but did not directly affect the security of the United States. As a general principle of international affairs, when there is an asymmetry of interest between two potentially colliding countries, the one more intensely involved in the outcome is likely to win, while the other side is more likely to swerve.

Another tactic for winning Car Chicken: Drive a large and imposing vehicle. If an armored cement truck is confronting a VW beetle in a Game of Chicken, which one is likely to swerve? This goes a long way toward explaining why the Soviets swerved during the Cuban Missile Crisis, since at the time, the United States was militarily superior to the USSR, both in terms of conventional forces available in the Caribbean, as well as nuclear might. Following that near-disaster in 1962, a high-ranking Soviet official was heard to say "You bastards won't be able to do this to us again," and shortly thereafter, the USSR began a major increase in its production of intercontinental ballistic missiles. Sadly, the resulting nuclear arms race didn't enhance international security. Although there may be a certain kind of stability when an armor-plated SUV bears down on a tiny sports car (because the latter is likely to swerve), everyone becomes, if anything, *less* safe when two monster trucks confront each other.

Even an asymmetry in "vehicle size" doesn't always enable us to predict the outcome of Chicken. In late 2002 and early 2003, the Bush administration played Chicken with the regime in Baghdad, demanding that Iraq comply with U.S. demands, or face war. The United States military was incomparably stronger than Iraq's, so the result of such a collision was never in doubt. Here was a tractor-trailer bearing down on a bicycle. Why didn't Saddam swerve?

Maybe he didn't believe that the United States really meant it, and was convinced that at the last minute, Bush and company would turn aside. Maybe he clung to the hope that international opinion would intervene and grab the steering wheel of U.S. foreign and military policy, forcing it to swerve despite the driver's intent. Maybe Saddam couldn't tell which way he was supposed to swerve; after all, since U.S. demands—for eliminating "weapons of mass destruction," for renouncing any connection with al-Qaeda, for "regime change"—kept changing, he might well have concluded that however he swerved, the United States would nonetheless swerve into him. In short, maybe the Iraq War wasn't a Game of Chicken after all, but rather a case of profound asymmetry, in which one side (for a variety of reasons) wanted a collision, which the other simply couldn't avoid.

In any event, an avid Chicken player has other options. For example, she might consider trying to reduce the Punishment she would suffer as a result of defecting. Not only would this improve her chances of surviving, it could also increase the chances that the other player—knowing the situation—would swerve. All things being equal, the likely victor in a game of Car Chicken is the fellow who shows up wearing Kevlar body armor and a well-designed crash helmet. The bomb-shelter craze of the 1950s was motivated, in large part, by strategists' worries that if the Soviets had an advantage in "civil defense" (which actually was neither civil nor defense), they would be emboldened to take more competitive risks vis-à-vis the United States. So Americans were encouraged to build their own family fallout shelters, less in hope of saving lives in the aftermath of a nuclear war (think crash helmets in a Game of Car Chicken) than as a policy instrument, hoping to stiffen the national spine in the event of a nuclear confrontation. And, of course, in the hope of making the Soviets more likely to swerve.

With the Cold War behind us, there is something arcane and bizarre about these parallels between games of Car Chicken and the Cuban

Missile Crisis, something that evokes nervous laughter when one considers the absurd logic of competitive risk taking. Keep laughing, but do so warily, since Games of Chicken have a habit of creeping up on all of us. Better yet, consider the advice offered by a top-secret Defense Department computer in the 1980s movie, *War Games*: "The only way to win is not to play the game."

Ultimatum Games

The Ultimatum Game is a great favorite among game theory fans. It isn't usually considered a Game of Chicken, but in a sense, it is one. As usually envisioned, Chicken requires that the two players make their decisions simultaneously, and independent of each other; in the Ultimatum Game, as we shall see, moves are sequential: one player moves, then the other moves in response. But if the moves are irrevocable, as they are in the Ultimatum Game, then Chicken is the logical outcome.

In fact, Ultimatum may actually represent the most common form of Chicken, in which head-on collisions frequently occur, but with consequences that aren't terribly severe. This is probably no coincidence: Although automobiles headed toward each other certainly get our attention, they don't happen very often. Most of the time, the vehicles in question are like old-fashioned bumper cars, colliding frequently, although without much impact or damage.

Here's how the Ultimatum Game works. Imagine that there are two people, one of whom—the "divider"—has just been given $10. His job is to divide up the money with someone else, the "decider." Under the terms of the Ultimatum Game, the divider proposes a division of the money (say, $7 for himself and $3 for the decider), whereupon it is up to the decider whether to accept the offer, or reject it. If accepted, the money is divided as agreed. If rejected, no one gets anything, and divider and decider go home empty-handed.

Each player, of course, brings a different perspective to the Ultimatum Game: The divider wants to get as much money for himself as possible, but he has to worry that if he proposes a very unfair distribution, the decider may refuse the deal, in which case no one gets anything. So the divider is tempted to be greedy, but not too greedy. For his part, the decider also wants to get as much money as possible, but because of the game's rules, her only logical play is to accept whatever the divider proposes. For example, even if the divider proposes $9 for herself and only $1 for the decider, the latter should accept, because disappointing as it might be, even $1 is better than nothing. But at the same time, the divider can't count on the decider being reasonable.

There have been numerous laboratory studies of the Ultimatum Game, typically involving individuals who do not know each other, and who have no reason to expect that they will interact in the future. Hence, they should be concerned only with maximizing their return from the game itself. Interestingly, the most frequent divider's offer is 50:50, perhaps because dividers often don't understand that they could obtain a more favorable payoff if they were more selfish. Also noteworthy: Deciders often reject offers giving them less than 30 percent, perhaps because they, too, don't really understand the logic of the game.[10]

Or alternatively, maybe they do. Maybe dividers are more clever—about human nature—than one might think. Perhaps they intuit that deciders may well prefer getting nothing to dividing the "pie" unfairly. And so they generally offer an illogical, but fairer division. Maybe, for their part, deciders typically turn down unfair offers because human beings are deeply imbued with a sense of justice and a refusal to be exploited, even if insisting on equity means getting less payoff in a given encounter. In the long run, over our evolutionary history, perhaps individuals who accepted an unfair division—of rocks, animal skins, mates, and so forth—were more likely to be taken advantage of in future exchanges. If so, it would actually be more logical to say no to an offer that made one look like a Sucker, even though getting nothing instead of something seems like a Sucker's payoff indeed.

Interestingly, deciders in Third World countries routinely reject unbalanced offers even when this means turning down the equivalent of several months' wages! Ask yourself how you would feel if the game were being played for, say, $10,000. You are the decider, and you've been offered $10, with the divider wanting to pocket $9,990. How about $3,500 versus $6,500?

The Ultimatum Game differs from most Games of Chicken not only in that the moves are sequential, but also because the players do not have symmetrical power. One is dividing, the other deciding. The classic case, Car Chicken, *is* symmetrical, but symmetry isn't the key to whether an interaction is a Game of Chicken. The relationship of the payoffs is. Here is the Ultimatum Game as a 2×2 matrix:

		Decider	
		accepts offer	rejects offer
Divider	fair division	R, R	−, −
	unfair division	T, S	P, P

For the divider, proposing an unfair division is equivalent to defecting; that is, selfishly going straight ahead toward an unbalanced division. Proposing a fair division is equivalent to cooperating, agreeing to swerve. For the decider, accepting an unfair offer is equivalent to swerving when the other player defects (i.e., a Sucker's payoff of a sort), while to reject an unfair offer is to defect in response to the other's defection, whereupon the divider and decider collide and both receive the lowest payoff, in this case none at all.

The divider is therefore tempted to defect—by proposing an unequal division—in hopes of getting the highest payoff (T). At the same time, the divider is likely to be satisfied with the medium-high payoff, R, of mutual cooperation (a fair division), and will be most disappointed with P, the Punishment that arises if the decider smashes the pie instead of accepting what the divider has proposed. The Sucker's payoff, S, goes to the decider who elects to accept a divider's unfair proposal. It is a poor return, but not as bad as the payoff both get if the pie is destroyed instead of divvied up. (For consistency, the upper-right corner in the matrix should be "S,T," but instead it is blank, because there is no real Temptation for the decider to reject a fair offer.)

In any event, the payoffs in most Ultimatum Games are ranked T > R > S > P, making it a disguised Game of Chicken, albeit one that is less damaging and less exciting than two automobiles hurtling toward each other.

In real life, some ultimatums are intended to be rejected, typically when the "divider" is actually playing to an audience and seeking to make a show of cooperativeness. Others are declined because the cost of accepting (of swerving) may be too great, especially when others are watching (lost credibility, reputation, etc.). Sometimes, the cost of being derided as a Chicken is greater than that of a collision.*

Ultimatum Games may be especially important because they demonstrate that Games of Chicken are probably more widespread than many of us realize. It is easy to be blinded by the drama of an imminent car crash. But we would probably be wise to look for lower intensity Games of Chicken in our daily lives. Prick up your game theoretic ears, for example, whenever someone mutters (or you find yourself thinking) "I wouldn't give him/her the satisfaction," or "Don't make me do this" or "This hurts me even more than it hurts you." And be alert to whether a kind of collision occurs, after which—as with most collisions—both parties are worse off.

Social Chickens

We've seen that Prisoner's Dilemma has a multiperson aspect: the social dilemma, in which every individual does best by defecting, but society, taken as a whole, does worse. Similarly, there is a multiperson version of the Game of Chicken. (After all, most chickens live in flocks.) William Poundstone calls it a "Volunteer's Dilemma," in which everyone wants someone to do something—to volunteer—but no one wants to be the one who does it. I call it Belling the Cat.

		Society's perspective	
		someone does it	no one does it
Your perspective	do it	3	2
	don't do it	4	1

In the original "belling the cat" story, a group of persecuted mice get together, announcing that their patience has run out: the time has come

---

*It seems clear that obsession with saving face is especially intense among human beings, just as it is within certain human cultures as opposed to others. Animals, too, are very much taken with matters of social status, although I do not know of any cases comparable to the human Ultimatum Game, in which animals refuse an offer because of not wanting to appear wimpy.

to do something about the mouse-eating proclivities of the local feline. The mice come up with this brilliant suggestion: Let's affix a bell to the cat, so it can no longer sneak up on its potential victims. Fine, but *who* bells the cat? If you are a mouse, you'd dearly love someone to volunteer for the job; mouse society as a whole would certainly be better off. But each individual would be best served if another individual did the job, whereas the worst of all worlds results if no one steps forward, with everyone delaying in hope that someone else will volunteer.

In Joseph Heller's novel *Catch-22,* Yossarian refuses to fly a suicide mission. His superior officer asks, "What if everybody felt that way?" Yossarian's response: "Then I'd certainly be a damned fool to feel any other way. Wouldn't I?"

Okay, we aren't mice. Nor are most people expected to volunteer for suicide missions, or even ordered to do so. But neither are we immune to the Volunteer's Dilemma. For example, take the Vaccination Game.

There is no doubt that vaccination is a public health benefit; in fact, an enormous one. Everyone is better off if everyone is vaccinated. But, in fact, there is a small risk associated with being vaccinated, since the vaccine itself could either produce the disease or generate an undesired side effect. After all, vaccines consist of inactivated disease-causing agents, or closely related pathogens, found in other species; occasionally, vaccines are somewhat more active than intended. This explains the reluctance of federal authorities to authorize widespread smallpox vaccination, even in the aftermath of the post–September 11 anthrax scare. It is also true that the higher the vaccination rate, the less likely that you will be infected with the disease, whether or not you have been vaccinated. So, from your selfish perspective, the best situation would be for you—and your children—to remain unvaccinated, so long as everyone else doesn't adopt the same strategy.

Some people do refuse vaccination, even when it is mandated by law. They can also be heard, on occasion, announcing that they don't understand what the fuss is all about: after all, *they* haven't been vaccinated, and they've never contracted measles, or whooping cough, or diphtheria. Unmentioned in such cases is that the reason they haven't been afflicted is that just about everyone else *has* been vaccinated:

	Everyone else	
	gets vaccinated	doesn't
get vaccinated	3, 3 (R, R)	2, 4 (S, T)
don't	4, 2 (T, S)	1, 1 (P, P)

You appears at the left, spanning the two bottom rows ("get vaccinated" and "don't").

In short, you would be best served if everyone else "volunteers," using their bodies as a communal bulwark to protect you against the disease in question. At the same time, most people agree to be vaccinated so long as everyone else does the same (that is, all parties cooperate; the equivalent of swerving in a Game of Chicken). The worst outcome arises if no one gets vaccinated so that the risk of disease is greatest (both sides defect; the equivalent of colliding).

A major scandal has been simmering in the United Kingdom, where the National Health Service provides a "triple jab," effective against measles, mumps, and rubella. But a research study has linked it— whether accurately or not—with childhood autism, a finding that has become widely known. Many parents have therefore been refusing to allow their children to be inoculated, and there is evidence that the frequency of all three diseases is beginning to increase as a result.[11]

At first glance, the Vaccination Game might seem to be a social dilemma; that is, a Prisoner's Dilemma in which each individual plays against the group. After all, the highest payoff for each individual comes when he or she avoids vaccination (getting T, the Temptation to defect) while everyone else participates (and gets S, the Sucker's payoff). And the next highest payoff comes when everyone gets vaccinated (R, the Reward of mutual cooperation). But the Vaccination Game is truly a case of social chicken, since the lowest payoff isn't the Sucker's S—obtained if you are vaccinated while no one else is (since that way, you probably avoid the illness and suffer what is in any event a very small risk)—but rather, when *no one* gets vaccinated, in which case you are at unacceptable risk of contracting the disease and thus suffering P, the Punishment of mutual defection . . . or, in this case, the Probability of Protracted Patienthood. As is everyone else.

There is a "free-rider" component at work here as well, since, as with the social dilemmas described earlier, the nonvaccinator is seeking to take advantage of the (modest) sacrifice of everyone else. In such cases, the "free rider" is immoral, although rationally pursuing his or her self-interest. This is why governments often pass legislation requiring vaccination, but also why noncompliance will probably always be a fact of life.

War and Peace

Game theory is at its worst in matters of war and peace. Not that it is especially inapplicable to such situations; rather, it has been troublesome in its consequences, precisely because it is so relevant. It was extremely popular among strategic analysts who were hired by the Defense Department in the early days of the Cold War, and whose hawkish preferences generally preceded their interest in the mathematics of game theory. But there is more to it than that. To some degree, game theory *generates* hawkishness, especially among people immersed in Games of Chicken. We have already seen that iterated games of Prisoner's Dilemma lend themselves to a self-fulfilling prophecy of mutual defection. When it comes to Games of Chicken, things are even worse.

Thus, Chicken leads to a bias toward sending messages of threat and determination. (It may also be that people with a penchant for threat and determination are especially prone to playing Chicken.) But it also seems likely that once people see the world in game-theoretic terms, their perspective is narrowed as a result. If all you have is a hammer, goes the saying, everything begins to look like a nail.

The late 1940s and early 1950s were a time of particularly widespread speculation about the desirability of "preventive war" against the Soviet Union. It was also the heyday of early game theory, most of it centered around mathematicians and economists at the newly established RAND Corporation (a creation of the United States Air Force; the word *RAND* was derived as an acronym for Research and Development). Seeing the post–World War II antagonism between the United States and the Soviet Union as a Prisoner's Dilemma, the RAND game theorists concluded in their mathematical wisdom that it was better to defect than run the risk of being suckered. Even more seriously, better to defect quickly, when only the United States had nuclear weapons, rather than wait until the Soviet Union could build itself into a nuclear power in its own right, whereupon we—and the world—would be vulnerable to potentially omnicidal Games of Nuclear Chicken.

In short, the temptation to defect was strong, especially when the United States had a nuclear monopoly and the Soviets did not have the option of defecting in return. Let's be clear, and get away from the dangerously sanitizing language of game theory: To defect, in this context, meant to attack the other side with nuclear weapons, causing the deaths of tens of millions of human beings, probably even more. No wonder game theory and game theoreticians seemed too clever for their own good. Not to mention everyone else's.

Even the fervent antiwar and antinuclear spokesperson Bertrand Russell briefly entertained just this idea. Notably, after things equalized—especially when the Soviets obtained hydrogen bombs and warheads—talk of preventive war quickly died out. At that point, the situation more closely resembled a Game of Chicken, and, ironically, a kind of stability ensued. (Not coincidentally, the George W. Bush administration, which recently revived the doctrine of preventive war, directed it specifically at countries—notably Iraq—that lacked the ability to retaliate, while treating others—such as North Korea and Iran—much more gingerly.)

Game theory has long held itself up as a technique for rational decision making. Indeed, much of the criticism it has received focuses on its cold, seemingly bloodless mental processing of options ("scenarios," in nuke-speak). Yet at the same time that game theory and its practitioners have typically occupied the hawkish, promilitary end of the political spectrum, those of the dovish, propeace persuasion have typically complained that much of the momentum for arms races and distrust of the enemy has derived from fundamentally irrational mental processes, notably hatred, fear, intolerance, and greed (the latter on the part of arms manufacturers). What the world needs now, according to many antinuclear activists, is not just love, but clear-headed thinking about the danger of nuclear war.

In fact, game theory is especially challenged by irrational actors, who will get their due in the concluding chapter of this book. The Strangelovian mind-set has long been a source of amazement and aggravation to peace activists, who have some difficulty with any supposedly rational calculus in which millions of lives are tossed about like so many bits of confetti. There is also the paradox of assuming—as nuclear strategists typically do—that it is possible, even expedient, to scare the pants off one's opponent with ostensibly credible threats of immediate incineration, and then expect the recipient to behave with exquisitely calibrated rationality.

Add to this the problem of playing Chicken, on occasion, with an avowedly irrational opponent. As we have already seen, one of the acknowledged strategies for winning a Game of Chicken is to feign lunacy. What about those who are really crazy (at least by the standards of whoever is driving the other car)? Does the lunatic automatically win? Consider, for example, terrorists who are not only potentially death dealing, but also death seeking: they are not readily deterred. After all, deterrence relies on the threat of unacceptable retaliation. A simple payoff matrix gives a would-be attacker two options: attack or refrain. It also gives the victim two options: retaliate (or, at least, possess the capacity to retaliate), or don't. The point of deterrence is that so long as a potential victim has the capacity to retaliate in a way that would be unacceptable to the attacker, said attacker would refrain from attacking. Simple enough.

But what about an attacker who is not averse to suffering retaliation, or even certain death? Game theory calculations based on a seemingly rational concept of self-interest do not get us very far when it comes to dealing with people who value their conception of the afterlife more than their present existence. Deterrence, which often involves threats and Games of Chicken, loses its potency if the other player isn't averse to a crash. Worse yet, if he is actually looking forward to one.

Of course, even this problem is not necessarily fatal—to game theory, that is!—since in principle, one need only adjust the payoff matrix accordingly, after which even the looniest suicidal terrorist can be revealed as a rational payoff maximizer. Some Islamic suicide bombers, for example, evidently were convinced that in return for their "heroic"

death, they would enjoy an eternity in paradise, experiencing volup-
tuous meals and surrounded by literally dozens of beautiful virgins,
eager to cater to their sexual desires.

Given this expectation, is it "unreasonable" for suicide bombers
and other terrorists to act as they do? And what will deter them?

Peace researcher Lewis Richardson, a Quaker who developed a
detailed mathematical model for arms races, believed that his theory
described

> what people would do if they did not stop to think. Why are so many
> nations reluctantly but steadily increasing their armaments as if they were
> mechanically compelled to do so? Because . . . they follow their tradi-
> tions . . . and instincts . . . and because they have not yet made a suffi-
> ciently strenuous intellectual and moral effort to control the situation.[12]

By contrast, military historians and scholars of a more right-wing
persuasion claim that the decision to participate in an arms race or
even to go to war is often highly rational. Game theoretician and
Nobelist in economics John Harsanyi* considered Richardson's view
to be "oversimplified and dangerous," and that

> it leads to serious underestimation of the political forces actually gener-
> ating and maintaining international conflicts; and in particular it pre-
> cludes understanding of the fact that . . . non-cooperative behavior may
> often be the only rational and realistic response to a given international
> situation.[13]

The Richardson view suggests that conflict is largely a result of irrationality, and that it can be diminished by better information and a
more enlightened view—empathy—of one's opponent. (To be distinguished from *sympathy,* which implies wanting to help the other; *empathy* simply involves understanding, possessing a good sense of what is
going on inside someone else.) By contrast, the game theory view says

---

*Who shared the 1994 Nobel Prize in economics with John Nash, and Reinhard Selten, he of the
"trembling hand."

that it is wishful thinking, or worse, to imagine that conflict can be prevented. Thus, game theory is offered as a tool to help manage conflicts, not eliminate or avoid them.

My suggestion is more modest. I propose that game theory is best deployed—indeed, perhaps the only way it can be usefully deployed—in helping us conceptualize conflict. I wouldn't look to it for advice, just for a bracingly clear view of what's going on. Often, it's a Game of Chicken.

Take the war in Kosovo. The Serbs felt that they had two choices: attack the Kosovars (defect) or refrain from doing so (cooperate). NATO also saw itself as having two choices: attack Serbia (defect) or do nothing (cooperate). What happened? Both sides defected, and both collided, with the Serbs attacking Kosovo and NATO attacking the Serbs.

Similarly, in the weeks preceding World War I, Austria had to decide whether to attack Serbia in retaliation for the assassination of the archduke (and, mainly, to retain credibility as a great power), while Germany had to decide whether or not to support its Austrian ally. Austria attacked, and Germany supported. At the same time, Russia and France had to decide whether to back Serbia (to which Russia was allied by ethnic similarity, with France allied to Russia by treaty). No one swerved, and World War I began.

Chicken is a very dangerous game because if either player seeks the highest possible outcome, he harms the other player (by moving away from the Pareto point, that precarious, vulnerable equilibrium), in the process subjecting both himself and the other player to potential disaster. Game theory may do everyone a service insofar as it alerts us to the risks of such "statesmanship" gone awry. But it also offers a costly disservice when its relentless logic persuades practitioners to foreclose other options. World War I, and probably the Kosovo War as well, could doubtless have been averted if world leaders were open to alternative moves, and less wedded to risky strategies.

Which leads us, at last, to *brinkmanship*, a term closely associated with the nuclear age policies of John Foster Dulles, Dwight D. Eisenhower's secretary of state and obsessive Cold Warrior.* The theoretical

---

*By the way, don't confuse brinkmanship with its nonexistent but oft-cited cousin, "brinksmanship," which presumably refers to the robbing of armed cars.

underpinnings of brinkmanship were, in fact, largely developed by economist-turned-nuclear-strategist Thomas C. Schelling, who was influenced by—and, in turn, exerted substantial influence on—game theory. Schelling emphasized that one way for dangerous potential antagonists to play Chicken (including, but not limited to, Nuclear Chicken), is for each side to allow the risk of a collision to escalate, up to the point at which one participant or the other can stand the risk no longer. Then, the one revealed to have the least tolerance for such a game or the weakest nerves or the greatest regard for human life, loses (i.e., swerves). As a result, each side pushes the other as close to the brink of disaster as possible. In 1956, following one of many U.S.-Soviet Cold War confrontations, Dulles was quoted in *Life* magazine as proclaiming, "Of course we were brought to the verge of war. . . . If you try to run away from it, if you are scared to go to the brink, you are lost. . . . We walked to the brink and we looked it in the face."

Why did the Chicken cross the road? To get away from those who insisted on playing such stupid, dangerous games.

When played between two roommates at risk of running out of breakfast cereal, brinkmanship is unlikely to be lethal (and, hence, the brink is that much more likely to be crossed). Maybe the closest approximation to a chickenlike collision between two recalcitrant roommates is . . . running out of toilet paper. But when the capacity for killing millions, or even ending life on earth, is at hand, the "game" becomes more than a little intimidating, not to say unwise. Fried chicken, anyone?

The Conscious Chicken

One of the interesting puzzles of evolution is why people have developed such large brains. Looked at one way, it isn't a puzzle at all: We have big brains because there is a distinct evolutionary advantage to being smart. Among our early primate ancestors, the ones with bigger brains were more successful—and, therefore, left more surviving offspring—than the pinheads. Selection favored increased intelligence, and this, in turn, produced individuals who were more intelligent than their fellows.

But this doesn't solve the mystery. Although biologists agree that natural selection favored intelligence, it remains unclear exactly how or why this took place. Presumably, it wasn't that individuals who were especially good at solving differential equations left more descendants than their competitors. More likely, there was a particular advantage in figuring out how to use primitive tools, perhaps to dig for food, construct shelters and weapons, or how to communicate more effectively, and, therefore, to raise successful offspring, to organize raiding or defensive parties, or maybe coordinate a hunt, or pull together one helluva party. Another interesting suggestion is that maybe our large brain, with its capacity for playful and seemingly useless and excess creativity, may have evolved as a way of attracting mates, the hominid equivalent of a peacock's tail.[14]

Biologists and anthropologists have also speculated that the remarkably rapid evolution of our brainy intelligence may have been intimately connected with the importance of complex, sophisticated interactions among other primitive human beings. After all, there is nothing unique about having to face potential predators, deal with life-threatening climatic events, or achieve the best possible mating and child-rearing arrangements. Most mammals—indeed, nearly all animals—have to make similar adjustments. Why should human beings be so special? One possibility is that the advantage of being smart may have been particularly intense when it comes to navigating a world of interactions among other individuals, each of whom is also smart. Maybe intelligence became especially rewarding, therefore, in an environment of competing intelligences, in which the greater the competition, the greater the payoff. And furthermore, the greater the payoff, the greater the selective advantage of increased intelligence. If so, then the advantage of having a large brain and an intellect to match may have become a kind of positive feedback system (or "vicious circle," to use a more negative phrase), in which the benefit of brains increased as others, in turn, were also brainy.

In the next chapter, we'll look into some wonderful situations in which animals—even those with unimpressive noggins—are remarkably good game players. Intelligence, it turns out, is not a prerequisite for effective strategizing; indeed, it occasionally gets in the way! But it is at least possible that under some conditions—notably, when playing games against others who are also especially good strategists—there might be an additional payoff to being flexible and imaginative in one's responses. If so, given that animals, too, have games to play, why aren't they as smart as we are? For one thing, it seems likely that their games simply aren't as sophisticated as ours; for another, keep in your brain the thought that brains themselves are expensive organs, using enormous amounts of energy and being quite vulnerable to injury. With such drawbacks, it wouldn't be surprising if big-time brains evolved only in situations of big-time game playing; that is, when the other players also bring substantial intelligence to the game so that the cost of building those demanding cerebral hemispheres is compensated by the benefit of having them.

Insofar as our mental evolution was strongly influenced by our ability to play games, why stop with intelligence? Perhaps game theory might explain human consciousness as well. In a sense, computers are highly intelligent; after all, they can play winning chess and perform very difficult calculations, but they don't show any signs of possessing an independent and potentially even rebellious self-awareness, like HAL, in *2001: A Space Odyssey*.

It may be that consciousness is the sine qua non of humanity, but wouldn't plain old intelligence be enough? Perhaps consciousness arose because it creates a degree of self-awareness. And why is self-awareness so useful? Here we are, at last: Maybe a self-aware individual is better able to anticipate what another individual is likely to be thinking and feeling and, therefore, doing. This would be particularly valuable if that other individual is bearing down on you, playing Chicken.

A gamey environment gives a special advantage to individuals who do not simply act or react to their environment, or even to the actions of others, but rather, who can mentally step outside their own heads and enter—hypothetically—the mental space of their colleagues, opponents, and fellow players. This is because the ability to coordinate with others, to modify one's behavior as a function of what the other player is likely to do—in short, to take the other's behavior into account—is at the heart of being a good gamer. More to the point, when dealing with others who are themselves intelligent, a degree of intellectual empathy may well be crucial to success in any of the complex games of life: "If I do this, he might do that; if he does X, I'd better do Y . . ." Or even: "He's likely to do this with a certain probability, so I'd better do this with a corresponding probability." A good game player should not only keep track of what other players have done previously—since this has implications for what these others will likely do in the future—but should also be attuned to what he or she has done in the past, since this is likely to influence how one is perceived by others, and how those others are likely to respond. A calculating kind of empathy thus seems crucially important for every game player, insofar as one's opponents are similarly endowed. And game playing, in turn, is so important that we may owe not only our intelligence but even our consciousness to it.

Self-consciousness may therefore also have a special role to play. After all, when payoffs depend, in part, on how the other player behaves, and when such actions are themselves influenced by what you have done and are likely to do, there is much to be said for being able to do as the poet Robert Burns advised: "to see ourselves as others see us." And given that Games of Chicken are so powerfully imbued with conse- quence for social and biological success, it would not be at all surpris- ing if consciousness and Chicken are inextricably connected, whereby the more acute the consciousness, the better the Chicken playing.

As ever in this enterprise, there are wheels within wheels. For example, although knowledge is generally preferable to ignorance, sometimes it is possible to know too much. Even in Games of Chicken. For example, if player A knows what player B is going to do and—especially—if B knows that A knows, this can actually be advantageous to B and para- doxically troublesome for A. If A knows, for instance, that B is not going to swerve, then this extra information actually confronts individ- ual A with a problem; he has no choice but to swerve, and thus, to lose. (It may be some comfort to know that in doing so, he has avoided an even bigger loss, if both players had gone unswervingly to their doom.) So, herein lies another paradox: There may be limits to how smart one ought to be! Maybe, indeed, this is one reason we are as smart as we are, and not smarter.

In a sense, being omniscient is equivalent to moving second instead of simultaneously with the other player. In certain games, such as one-time Prisoner's Dilemma, such knowledge needn't make much difference, since the best move is to defect, regardless. But when playing Chicken, you want to defect only when you can be confident that the other guy is *not* doing the same thing. So in this case, counter-intuitively, it is good to know something, but not too much.

And this, in turn, brings us to situations in which game players might not consciously know anything at all, but are often pretty good strategists nonetheless.

6

Animal Antics

By this point, it should be clear that humans don't have a monopoly on game playing, and that game theory, for all its vaunted logic and rationality, applies to other living things, too. Intellectually challenged creatures could hardly be expected to work out the math, or the logical reasoning involved. But, in fact, game theory works best among the "lower" forms of life.

The key is to recognize that payoffs in traditional human games (time saved, calories gained, competitive situation improved, etc.) can be equally important to animals, or even plants; if anything, they are likely to be more significant than among people. And, moreover, natural payoffs are translated directly into more players, each likely to be employing those strategies that generated the strategies in question. This is because in the natural world, these payoffs aren't just a matter of preference but a direct contributor to evolutionary success. Higher payoffs can mean more food, more safety, and more mates, all of which ultimately mean more successful reproduction, and therefore, more individuals in the next generation who are inclined to make high-payoff "moves" of the same sort. In a typical game-playing experiment, people are rewarded with money, or tokens, or something of that sort; in the game of life, organisms are rewarded with reproductive payoffs that are considerably more real, and that result in winners eventually replacing losers.

Evolution can tell us a great deal about game theory, especially in one particular domain: more profitable strategies become increasingly abundant over time, while less successful ones diminish in frequency. Or put it this way: The best moves from the past are more likely to recur in the future. Or, even more strongly: The likelihood of a particular strategy being employed varies directly with its payoff thus far. At the same time, game theory can tell us a great deal about evolution; notably, it illuminates nearly every biological event in which "fitness"—success in projecting genes into the future—depends on the actions of other organisms. The result is a lot of light, in all directions.

When it comes to evolution, the bottom line is that the success of a particular strategy isn't simply a matter of accumulating payoff points, but rather, of gaining "fitness," which is to say, projecting one's genes into succeeding generations. (Or, as "selfish gene" theorists hold, genes giving themselves a boost, by benefiting exact copies that reside in other bodies.)

Evolutionary theory describes what actually happens but not what should happen. By contrast, game theory describes what should happen, if all players are strictly rational, but says nothing about what actually does occur. The pleasing symmetry of these two statements is, however, a bit overstated. This is because evolution also says what should happen, but not in a moral sense; rather, it depicts what should happen if individuals are behaving so as to maximize their fitness, which is equivalent to maximizing their payoffs.

In the world of biology—which is to say, the world of living organisms, including, but not limited to, *Homo sapiens*—payoffs have meaning insofar as they contribute to evolutionary success. If one payoff is higher than another (more calories, more mates, more offspring, etc.) this means that, ultimately, it contributes more to the success and future representation of whoever or whatever is getting that payoff. In essence, evolution creates entities that are, if anything, *more* adroit at making decisions consistent with the rules of game theory than are human beings, whose brains readily intervene, generating responses that may be less automatic, and, sometimes, less efficient and more error prone.

"Between the idea and the reality / Between the motion and the act / Falls the Shadow," wrote T. S. Eliot in *The Hollow Men*. The human brain is a kind of shadow, falling in the space of animal automaticity, introducing the prospect of deep thought, inhibition, delay, and reconsideration. The mathematically precise structure of game playing is therefore more closely approximated by unthinking beasts—and even plants—than by intellectually sophisticated human beings who, by contrast, labor (or luxuriate) beneath their cerebral shadows.

The logic of evolution is logical indeed. Among the many behavioral tactics and strategies that were "tried" in the past, some worked better than others. Animals and plants never asked why, and never bothered (or were able) to figure out the details. Instead, those behaviors that conveyed the highest payoff prospered; those that didn't, didn't. The result is a direct correspondence between play and return, unencumbered by morality, second thoughts, emotion, or even an intellect that struggles to keep up with the options. No shadows. As suggested, human intelligence and consciousness may have developed because playing games with intelligent opponents put a special premium on intelligence and consciousness. In a world of morons, without any payoff to make up for the costliness of fancy but vulnerable gray matter, evolution would prefer decision making based on automaticity and rigid rule following.

Game theory applies more smoothly to evolution than to intellect for yet another reason: Payoffs have more literal, and meaningful, value when life and death are at stake. It is relatively easy to rank payoffs from most desired to least, but notoriously difficult to evaluate them numerically. Is a collision in a Game of Chicken twenty times worse than swerving and being a Chicken? A thousand times? On the other hand, when payoffs are measured in terms, say, of reproductive success, the numbers are direct and eerily accurate: If a particular behavior such as defending a territory in one location instead of another results in 1.23 times more offspring, then it is clear which is more successful, and by precisely how much.

The result is that animals play games. And as we shall see, they are very good at it.

Snaggle-Mouths, Side-Blotches, and Stability

The so-called Rift Valley lakes of Africa are home to hundreds of species of fish known as cichlids,* among which is a group known as *Perissodus*. Call them snaggle-mouthed cichlids, because each individual has his or her mouth twisted to one side or the other, either to the left or right. These bizarre twists of fortune are caused by some cold-blooded evolutionary game-playing logic: *Perissodus* fishes make their living by sneaking up on other fish and snatching a few scales out of their sides. The best way for these scale snatchers to approach their prey undetected is to surprise them directly from behind, but this angle of attack makes it difficult for a perpetrator to grab a bite out of the victim's tail. It might prove easier to approach the target meal from the side, at a right angle, but since the eyes of most targets are also located on either side of their heads, such a tactic wouldn't work so well. Instead, evolution has offered up a solution: The jaws of scale-snatching *Perissodus* are snaggled in one direction or the other. After sneaking up on their victim's tail, from directly behind, those cichlids whose mouths are twisted to the right, not surprisingly, attack the left side of their victims, and those twisted to the left attack on the right.

So far, so good . . . at least for the snaggle-mouths. The victims, on the other hand, can be expected to see things differently; in fact, they strain to do just this, to see any approaching *Perissodus* before they get nipped. Once bitten, they become especially shy, looking behind and over their shoulders (or whatever passes for shoulders among fish), from whence the snaggle-mouthed cichlids appear.[1] Now the plot thickens. Recall that because of the mechanics of scale nipping from behind, *Perissodus* come in two distinct types, twisted left or twisted right. When the left-twisted form becomes abundant, their victims—having been attacked primarily on their right flanks—become especially vigilant about additional attacks from that side. As the victims get particularly leery about being approached on their right flanks, left-twisted *Perissodus* are less successful in their hunting, and so their

*Pronounced "sick-lids," by the way, not "chick-lets."

numbers begin to decline. The result? Right-twisted *Perissodus* enjoy an advantage. Accordingly, they increase in abundance until their victims "wise up" and stop looking over their right shoulders, turning their attention instead to their left. Eventually, the system reaches equilibrium, with equal numbers of right- and left-twisted *Perissodus,* and with victims scarred equally, on both sides of their bodies.

The biologists who studied this system wanted to explain why two different forms of snaggle-mouthed cichlids coexisted, and they settled on a game theoretic answer that brings to mind a children's seesaw. Success by either form—snaggled-right or snaggled-left—carries the seeds of its own destruction, as victims start looking out for the abundant type, whereupon the other, less abundant type gains in frequency, only to suffer a decline in turn.

Evolutionary theory—and also common sense, for that matter—suggests that there is a best way of doing most things, including a best way of having your mouth snaggled if you are a *Perissodus* fish. But this isn't necessarily true, especially if your payoff depends on what others are doing, and if those "others" include individuals employing your strategy. All the more so if that strategy does well when few others employ it, and poorly when many do so. The result can be situations in which two or more different types—like right- and left-snaggled fishes—live together in what appears to be peaceful coexistence.

Mathematically inclined biologists have become very enthusiastic about such equilibrium systems, in which distinct forms of the same species appear to have evolved to coexist in a state of stability. Next, a case in which there are three different forms, all flourishing, and all kept in play by a very gamey pattern of interaction. It's a reptilian version of the Paper-Rock-Scissors Game.

The players are known as side-blotched lizards, *Uta stansburiana.* Among these animals, which are common in the American Southwest, there are three different kinds of males, genetically different from one another, and each readily distinguished by different styles of mating and social behavior. "Normally aggressive" individuals maintain a small territory containing one female; "highly aggressive" males keep a large territory with many females; and "sneakers," which are unusually small, do not defend any territory and are often mistaken for females. This deception allows them to sneak onto the territories of other males, to copulate with the females residing there.

When a normally aggressive side-blotch male encounters a highly aggressive one, the latter nearly always wins. You might expect this to be the end of the story: highly aggressive males would replace normally aggressive ones, and, as a result, *Uta stansburiana* would consist entirely of "highly aggressive" males. But don't forget the sneakers. Because the high-aggressive guys are so busy patrolling their large territories and bullying the normal-aggressives, they are especially vulnerable to sneakers who, being mistaken for females, get to hang out with the lady lizards and obtain more than their share of copulations. And this copulatory payoff is translated into a reproductive payoff, which, in turn, means more sneaky lizards in the next generation.

At the same time, the success of sneakers against high-aggressives does not carry over to their battles with normal-aggressives. This is because normal-aggressives keep small territories with only one female apiece. As a result, normal-aggressives—although defeated by high-aggressives—are able to repel sneakers, since the normals, being monogamous and keeping small territories, don't have their attention so scattered.

The result? Normals are defeated by high-aggressives; high-aggressives are defeated by sneakers; and sneakers are defeated by normals. And so, the population cycles from a high frequency of normals to one dominated by high-aggressives, to mostly sneakers, back to normals, and so on. It is essentially an evolutionary Paper-Rock-Scissors Game, although the side-blotched lizard population fluctuates among the three forms,[2] whereas in Paper-Rock-Scissors, three "strategies" are maintained simultaneously in equal proportion.

As with Paper-Rock-Scissors, just as with the strange case of the snaggle-mouthed cichlid fish, there is no "best strategy" for side-blotched lizards. Payoffs depend on what the others are doing, and the players are remarkably good at their games. It's important to emphasize once again that no special smarts are required in the world of biology, in which creatures do not need to calculate their payoff probabilities. Natural selection does it for them.

For an animal with even less gray matter than fish or lizard, consider a seemingly unsavory insect, the common yellow dung fly, studied in exquisite detail by British mathematical biologist Geoffrey Parker. Female dung flies are especially partial to nice, warm, moist, slightly steaming cowpies—that is, fresh ones—on which they deposit their eggs shortly after mating. Not surprisingly, male dung flies have a comparable preference for only the freshest cowpoops, not because of a particular culinary fastidiousness on their part, but because they go where the lady dung flies can be found. The older the cowpoop, the fewer the females; accordingly, older cowpoops have fewer males because in the world of dating dung flies, the boys go where the girls are. This poses an interesting question for each sexually motivated male dung fly: Where should he hang out, looking for his lady love? Bear in mind that every male wants the highest probability of encountering females, which would suggest only bothering with the freshest cowpoops, where the most females can be found. (Also bear in mind that such males need not really "want" anything, consciously; rather, those that make the best decisions are most likely to produce offspring, who are themselves likely to make comparable decisions because they are carrying their parent's genes.)

The dung-fly dilemma is as follows: Insofar as more females attract more males, an individual male at a particularly fresh cowpie would likely encounter lots of males as well as lots of females, and, in the resulting male-male competition, he might do less well than if he lurked about on older poops, where there were fewer females but also fewer males.

What, then, should you do if you were a male dung fly who has just located a gloriously fresh cowpoop? Stay with it, at least at first. But, as it ages, it becomes less attractive to females, fewer of whom are likely to arrive; at the same time, however, it is also becoming less attractive to other males, more of whom are likely to depart, giving you better odds with any females who happen to show up. It's a game. If enough of your fellow males leave early, you should hang around, because even though there would be fewer females arriving, there would be less competition from other males. But if everyone stays, you should leave, since fewer females would be arriving, and the competition would become, if anything, more fierce than ever. What actually happens? Parker found that the males distribute themselves so that each male does about equally well; that is, most leave early but some hang around, so that the average mating success, per male, is pretty much the same regardless of which cowpoop they haunt.[3]

People are pretty much like dung flies.* Just substitute a checkout line at a grocery store for a fresh cowpoop. If the line at one cashier is short, it attracts more customers, just as people tend to emigrate from excessively long lines. As a result, shoppers distribute themselves so that everyone has about the same wait. Ditto for cars on a multilane highway. Although exasperated drivers may be prepared to swear that their lane is always the slow-moving one, the reality is quite different.

Sex

Here's a good question: Why are there equal numbers of males and females? The not-so-good answer is that males make equal numbers of X and Y sperm. The problem is that this response simply displaces the original "why" question by one step. Why do males make equal numbers of X and Y sperm? After all, if there were an extra payoff to producing more males or females, presumably a system would have evolved that had males biasing their sperm making to favor one type or the other; or alternatively, females would have evolved the ability to discriminate against X- or Y-bearing sperm, or maybe to selectively abort embryos of the less-favored sex. In fact, equality of sex ratio is one of the great constants in the biological world, not just in human beings but in living things generally.

The great geneticist and statistician R. A. Fisher first drescribed this puzzle—and its answer—in a book published in 1930, which prefigured the development of evolutionary game theory.[4] It harkens back to arguments redolent of our good friends the snaggle-mouth cichlids. Here, in brief, is Fisher's analysis.

---

*Such a statement is, I suppose, leading with my chin, just begging a critical reviewer to take a poke at this absurd reductionism. So, I'll modify it: "People are pretty much like dung flies, in some ways."

Imagine that there is a population of animals (or people), in which, for some reason, the sex ratio is somewhat unbalanced, say, more males than females. Imagine further, for simplicity, that the creatures in question are monogamous, so that the mating unit in each case is one male and one female. If you were a reproducing member of this population, would you be better off—that is, more fit in the evolutionary sense—if your offspring were boys or girls? In the game theory sense, would your payoff be higher if your offspring were primarily of one sex or the other?

Remember, in our hypothetical example there are already more males than females. This means that under monogamy, at least some of these males are nonreproductive bachelors, whereas every female is likely to have found a mate. Under these conditions, a newly arrived male would face an uphill struggle when it comes to reproduction, since he would be competing with other males who have already been excluded from the mating game. A newly arrived female, by contrast, would have a much easier time; after all, since there are unmated males around, it probably won't be difficult to snag one of these frustrated fellows. Indeed, nonbreeding males are likely to be not only available, but downright eager to find a reproductive partner. So when the sex ratio favors males, better to be a female. (And vice versa: If there is an excess of females, better to be a male.)

From the perspective of adults about to reproduce, a higher payoff (higher fitness) can therefore be gained by producing extra females when that sex is in short supply. And, of course, the converse also holds: When there is an excess of females, reproducing adults would enjoy a higher fitness payoff if they produced relatively more males. (In this case, males would be experiencing less competition, so fewer of them would be left out in the reproductive cold.) The result is that the rarer sex, whichever it is, would always be produced. Whenever the sex ratio teeters out of balance, evolution—working via the reproductive strategy of its participants—therefore brings it back to a 50:50 equilibrium.

What we have been discussing is a simple coordination game, in which reproducing adults play against the larger population, their "moves" being to produce more boys or more girls. Here, then, is the Sex Ratio Game:

		The population consists of more	
		males	females
Breeders produce more	boys	−	+
	girls	+	−

It's a simple system: The payoff to breeding adults depends on what the population as a whole is doing. A positive payoff comes from producing boys whenever females predominate, and girls whenever males are in the majority. Incidentally, games of this sort—in which individuals play against the larger population—are probably more common than one-to-one competition, just as social dilemmas are probably more common than Prisoner's Dilemmas.

The Sex Ratio Game assumes monogamy: In order for a male to reproduce, he has to pair up, exclusively, with a female, and vice versa. What about polygyny, when a male maintains a harem of many females? (Or polyandry, the rare situation in which a female keeps a "harem" of males?) You might think that in either of these cases, the balance point would be shifted so that, instead of a 50:50 ratio, the Sex Ratio Game would equilibrate at a ratio that equaled the proportion of breeders. If, say, a species had a breeding system in which one male typically kept a harem of ten females, then the payoff to producing males and females might balance when there was, on average, one male in the population for every ten females. But interestingly, this isn't the case.

Imagine a hypothetical 1:10 harem situation. If each female bears one offspring per year (think of elk, or seals), you would calculate the annual fitness payoff for every female as one, and for every male as ten. If you were a parent in this population, about to play the Sex Ratio Game, what sex of offspring should you produce? In game theory terms, what "move" or "play" would you make: sons or daughters? Clearly, you'd be more fit—enjoy a higher payoff—if you produced sons, since the average son gets ten times the payoff of the average daughter.

Introduce one additional male and, assuming that he is as likely as the preexisting male to hold the harem, an average male has a fitness of five, which is still higher than that of the average female, which remains constant at one. At what point will it be a matter of indifference to parents whether they produce males or females? When there are equal numbers of males and females. The result is that the sex ratio in our imaginary population, which was originally biased toward females, would quickly become balanced toward 50:50, the point at which the payoff for producing sons and daughters is equal. This is true whether the breeding system is monogamous or a sexually unbalanced seraglio, as in our example of a 10:1 ratio.

To bring the point home, let's imagine the same hypothetical harem situation of one male mating with ten females, but in a population with equal numbers of each sex. In this case, every female gets to reproduce, whereas, for each harem-keeping male, there are nine excluded bachelors. Once again, each female has a reproductive payoff of one. The harem-keeping male on the other hand has a reproductive payoff of ten, while the nine bachelors have zero. But it is misleading to look at the payoffs of individual males; instead, we have to calculate the payoff to the average male, whose reproductive success is one, just like that of each female. (Ten for the harem master, plus nine zeros—for the bachelors—equals ten, divided across ten males, equals one.) As a result, even when the actual breeding system is wildly unbalanced, so long as there are equal numbers of males and females, every son and every daughter is equally valuable to the parents who produce them, since a small number of highly successful harem keepers is precisely counterbalanced by a large number of male reproductive failures.

However you slice it, whenever the proportion of males to females becomes unbalanced, the rarer sex automatically becomes more valuable. Therefore, natural selection favors parents who produce individuals of that sex, which in turn rights the unbalance.

And, in fact, this is the way of the natural world. Whether monogamy, polygyny, or polyandry, a 50:50 sex ratio is one of the great constants of life. Thanks to game theory, we understand why.

Social Dilemmas Among Natural Groups

Remember our old friend, the social dilemma? It arises when individuals find themselves playing Prisoner's Dilemma against a larger group instead of a single opponent; but either way, there is strong temptation to defect. The more biologists examine the social lives of animals, the more we are finding that many of them—and not just those sexually violent mallards we encountered in chapter 4—are subject to social dilemmas. Moreover, there is no reason to think that either victims or perpetrators ever see themselves as impaled on any ethical horns.

Acorn woodpeckers provide a case in point. They are noisy, busy birds of the American Southwest, colored black and white with cheerful clownish slashes of red. They live in large social groups, sometimes numbering in the dozens. Acorn woodpeckers are renowned food storers. Where these birds are abundant, their habitat is dotted with conspicuous "storage trees," typically oaks that have been perforated with hundreds, sometimes thousands, of small holes, into each of which an acorn or two has been stuffed. It takes time and effort for a woodpecker to locate an acorn, drill an appropriate hole, and then deposit its prize. But there is a payoff: When food is scarce, acorn woodpeckers excavate the morsels from their storage sites.

There is also a dilemma: What is to stop an acorn woodpecker from being a freeloader, failing to store acorns along with his or her fellows, but then feasting on the accumulated bounty nonetheless? A freeloading acorn woodpecker would seem to enjoy the same benefits as honest acorn storers, without paying the costs; but if everyone is a freeloader, everyone loses out. The following matrix shows the payoffs to an individual, depending on whether it is a freeloader or a storer, and what others in the population are doing.

		Most birds are	
		storers	freeloaders
Individual bird is a	storer	R	S
	freeloader	T	P

The highest payoff—the Temptation of loafing while everyone works—accrues to freeloaders (defectors) in a population of storers (cooperators), and the lowest—S, the Sucker's payoff—goes to a storer in a population of freeloaders. If everyone stores, then all enjoy the

Reward (R) of a full larder when times are tough, whereas if everyone gives in to the Temptation to defect—that is, if everyone is a free-loader—then everyone gets a low payoff, P, the Punishment of mutual defection. As always, T > R > P > S. It's a classic Prisoner's Dilemma, played against society.

Fortunately for them, acorn woodpeckers have not studied game theory, because although logic dictates that storing should be a Sucker's game, and thus, less fit from an evolutionary perspective, the reality is that acorn woodpeckers do, in fact, store acorns! Most likely, other things come into play. Maybe acorn woodpeckers watch one another, not like hawks, but like acorn woodpeckers, and ostracize those who don't do his or her share. Maybe there is a selfish benefit to storing acorns, such as being able to remember precisely which holes are full, which yields a payoff when it comes time to extract some acorns. Or perhaps something else is going on that we don't yet understand. But in any event, there is little doubt that animals also experience social dilemmas, and have had to develop strategies for overcoming them.

Another example of a social dilemma in the animal world was described by Richard Dawkins in his book *The Selfish Gene*. There are many ocean-going, fish-eating birds, such as terns and kittiwake gulls, that make long, arduous trips to sea, catching fish that they bring back to their cliff-side nests. Imagine that, initially, a population consists entirely of honest fisher birds, each working hard to catch his or her dinner. Such a system would be vulnerable to invasion by pirates, lazy troublemakers who refuse to do an honest day's work, and instead lurk around the cliffs, ready to ambush the honest fishers when they return with bulging beaks. This would be exactly equivalent to the acorn woodpecker's dilemma, replacing freeloader with pirate, and storer with fisher. Once again, every individual should be best off defecting—in this case, tempted to be a pirate and receive payoff 4—and worst off being a fisher in a world of pirates—that is, a Sucker who gets victimized and ends up with payoff 1. Similarly, an individual could be rewarded and do pretty well (payoff 3) as an honest fisher so long as everyone else is fishing, and would receive a poor and punishing payoff (2) if it is a pirate when everyone else is also piratic:

		Most seabirds are	
		fishers	pirates
A given seabird is	a fisher	R (3)	S (1)
	a pirate	T (4)	P (2)

What keeps these seabirds from all flying the skull and crossbones? Maybe, as with our supposition about acorn woodpeckers, there is some sort of policing on the part of "most individuals," keeping everyone on the straight and narrow. Maybe honest fishers get a selfish benefit from what appears cooperative: Like acorn woodpeckers who might gain useful information by storing their own acorns, maybe there is something to be said for catching your own fish. Or maybe if pirates become troublesomely abundant, honest fishers can retaliate by swallowing their catch before bringing it home. Maybe honest fisher birds don't let their daughters marry pirates.

Another possibility, as Dawkins suggested, is that fishers can indeed be invaded by pirates, who do well for a time, since they are surrounded by easy pickings, the animal equivalent of Blackbeard among a bevy of fat merchant ships. But as the honest fishers decline and the nasty, selfish pirates increase, pirates run into trouble, since as their numbers expand they are increasingly likely to run into one another. At this point, it might pay to be an honest fisher once again, especially if fishers can somehow protect one another, perhaps by traveling in convoys. If the payoffs are right, the system could reach an evolutionarily stable equilibrium, in which both fishers and pirates coexist.

If so, the necessary condition is simply this: The average fisher must get the same payoff as the average pirate. But this doesn't necessarily mean equal numbers of both strategists. In fact, unlike the Sex Ratio Game, it is very unlikely for things to balance out at 50:50. Equal numbers are the balance point for the Sex Ratio Game because every child represents a precisely equal success for one male and one female, so whenever reproduction takes place, the average male and the average female profit equally. But there is nothing in the Pirate Fisher Game that suggests the two moves are equally beneficial. Maybe fishers and pirates coexist when there are five fishers for every two pirates, or maybe a different ratio will turn out stable. The specific proportion depends on the payoffs.

For most games—animal or otherwise—the key to stability is that each strategy is equally successful across the population, even though such a balance might be achieved with an *imbalance* in the number of players using the strategy. If a strategy is so successful it drives others to extinction, the system has settled down to what is called a "pure strategy," with everyone doing pretty much the same thing, and not much call for game theory.

Ironically, it is just possible that social dilemmas among animals are resolved more readily than among human beings, since our species must also negotiate concepts like fairness and justice, and thus, such reproachful terms as *freeloader* and *pirate*. For animals, as for evolution, the only criterion is what works.

Hawks and Doves

When lions fight one another, they usually aren't as ferocious or lethal as when attacking their prey. Fish, when aggressively inclined, often seize one another by their leathery jaws; the ensuing tug-of-war is unlikely to cause the sort of damage as would be inflicted, for example, by a simple—but unlikely—bite in the flank. Bighorn sheep bash heads, striking with potentially lethal impact on the one part of their anatomy capable of safely withstanding such force. Rattlesnakes are not immune to their venom; a bitten rattler will die. But when rattlesnakes fight each other, they generally struggle to push the other onto his back, resulting in a curious posture that, incidentally, gave rise to the caduceus symbol of medicine, and that also provides a nonlethal method for settling disputes. The loser is rarely bitten; most of the time he just slithers away.

These and many other examples of animal restraint were exaggerated during much of the twentieth century, culminating in the assertion by animal behaviorist Konrad Lorenz that lethally armed animals never kill one another. It is now clear that such pronouncements are inaccurate: Wolves, lions, and even chimpanzees occasionally commit murder. Nonetheless, like most old saws, the claim of animal restraint has proven to have plenty of teeth. Animals with the capacity to wreak havoc on other members of their species frequently refrain from doing so.

For decades, biologists aware of this restraint considered it noteworthy—even fascinating—but not surprising. After all, evolution was long considered to operate "for the good of the species," a perspective that was reinforced by examples of animal benevolence or inhibited violence. "Of course animals often refrain from killing each other," it was said, "it's for the good of the species." No longer. It is now widely acknowledged that evolution does *not* operate for the good of species, but rather, via the disparities in success of different individuals and—better yet—of competing genes. (Imagine genes that were in the habit of sacrificing themselves for the good of their species. Now, add one gene that didn't worry about the species but, instead, about itself. That selfish gene would have an advantage and would be promoted in succeeding generations at the expense of the other-oriented genes around it.)

In fact, it is now understood that species are simply individuals—and their genes—considered in the aggregate, and that competition among individuals and among their genes constitutes the primary arena in which natural selection plays itself out.

Which leads us to a thorny problem. If evolution rewards the success of individuals and genes, rather than that of larger groups, then how are we to explain those wonderful, fascinating cases of restrained lethality? If not for the good of the species, why do some individuals rein themselves in? And why is it that, in many species, some individuals are inclined to be aggressive and even violent, whereas others are more pacific? If they would be more fit (that is, if they received a higher reproductive payoff) by being violence prone, then why isn't everyone on a hair-trigger? Or if nonviolence generates a higher payoff, why aren't animals models of Gandhian *satyagraha*?

It turns out that there are many explanations, most of which are nonexclusive. One of the most interesting, and relevant to game theory, was first proposed by British evolutionist John Maynard Smith and the American mathematician George Price in an article titled "The Logic of Animal Conflict."[5] Maynard Smith and Price saw the struggles of animals, the surprisingly high frequency of restraint, and the coexistence of both violent and nonviolent behaviors in the same population (without either driving the other to extinction) as being the product of animal game playing. Their article became the cornerstone by which game theory and the study of animal behavior were linked.

The classic example, first devised by Maynard Smith and Price, and the focus of their innovative research, is known as the Hawk-Dove Game.

Imagine that there are two different kinds of individuals: hawks and doves. They are the same in every way, except in how they respond to competition. (Also, they aren't necessarily birds.) Hawks threaten and, if need be, fight; doves are pacifists who avoid violent conflict. To start, picture a population made up entirely of doves. Imagine further that these animals are hungry, but, fortunately, food is scattered around their habitat. If one dove meets another and there is some food nearby, they divide it equally. No friction, no fighting. To make the model a bit more realistic, assume also that there is a small cost to each dove, since they waste a certain amount of time gesturing pacifically to each other.

Now introduce a hawk into this dovish Eden. Hawks are fighters, and they are not about to share food with anyone. So, when hawk meets dove, the hawk threatens and gesticulates and prepares to fight, while the dove immediately backs down. The result? The hawk gets all the food and the dove, none. However, when hawk meets hawk, neither backs down. They fight. Eventually, one of them wins, and to the victor goes the food. On average, each hawk in a hawk-hawk battle wins half the time, so each gets one half the food, just as when dove meets dove. The crucial difference in the two strategies is that for the hawks, there must first be a fight. Hawk-hawk encounters exact a toll on both participants, the cost of being fatigued or wounded, as well as the possibility of being killed.

Here is a payoff matrix for a simple Hawk-Dove Game. Since the game is symmetrical, the payoffs are shown only for the player at the left:

	Dove	Hawk
Dove	½ food, minus cost of displaying	0
Hawk	all the food	½ food, minus cost of fighting

The result is an interesting dynamic. Imagine the fate of a hawk introduced into a population consisting of all doves. The hawk would prosper, and therefore become more abundant; after all, compared to doves, each hawk would be consistently better nourished since when hawk meets dove, the former gets all the food. But, as time passes, and there are more and more hawks and relatively fewer and fewer doves, hawks begin to encounter other hawks. What happens then? First they threaten and act "macho," but neither one backs down. Those same hawks who did well in a world made up mostly of doves find themselves at a disadvantage as the proportion of hawks increases and hawk-hawk encounters become unavoidable. The doves, admittedly, don't do terribly well when meeting hawks, since they get none of the food, but at least they don't run the risk of receiving a negative payoff (the cost of fighting); moreover, when a dove is fortunate enough to meet another dove, each receives half the food—the same that hawks get after having encountered another hawk—without getting beaten up first! It seems reasonable to assume that the time wasted when a dove displays its dovishness to another dove is less costly than the consequences of fighting when a hawk meets another hawk.

Put it all together, and here's what happens. When doves are abundant, hawks prosper, enjoying easy pickings in a world of compliant pacifists. But as hawks become more abundant, they begin to suffer compared to doves. In their success lies the source for their downfall. At this point, doves find themselves at an advantage, whereupon they increase in frequency, only to open the door to invasion by hawks once more. In short, each type does well when it is rare, poorly when abundant. If the system is in equilibrium, then the average hawk will get the same payoff as the average dove. Another way of saying this: Hawks and doves can find that their strategies offer the same evolutionary payoff, which means that hawks and doves reproduce themselves at the same rate. At such an equilibrium, it is an equally good strategy to be a hawk or a dove; this is what biologists mean when they say that the hawk-dove system can be evolutionarily stable.

This doesn't imply that there will be equal numbers of hawks and doves, any more than there must be equal numbers of fishers and pirates among seabirds, or of storers and stealers among acorn woodpeckers. Rather, it simply means that at equilibrium, the average hawk does as well as the average dove. When this is reached, the proportions of hawks and doves—whatever they may be—are maintained, kept steady because this is the point at which the average hawk is doing as well as the average dove. The exact numbers, in turn, depend on the payoffs, which in turn depend on the payoff value of the resources in dispute as well as the cost of fighting (for hawks) and of displaying (for doves).

In the example developed by Maynard Smith and Price, the value of the resource was arbitrarily taken to be 10, the cost of fighting—a unique hawk-hawk expense—was 20, and the cost of wasted time when doves display peacefully to each other, 3. Given these values, it can be calculated that an equilibrium is reached when hawks make up 8/13 of the population, and doves 5/13. It could also be reached if each individual varied his or her behavior, in which case the stable state would be reached when each individual played hawk 8/13 of the time and dove 5/13. Mathematically, the results are the same, although biologically, they are quite different. The first implies that individuals are rigidly predetermined as to their strategies, something more likely for simple organisms such as invertebrates, whereas the latter assumes that they are more flexible, as would be especially true of relatively large-brained vertebrates such as birds and mammals.

There is nothing magical about the proportions of hawks and doves found at equilibrium; they depend entirely on the payoff values. In fact, there is something intuitively appealing about the dependence of the former on the latter. Consider this, for example: If the cost of fighting increases, the proportion of hawks at equilibrium goes down, which makes perfect sense. As hawk-hawk battles become more likely to result in death or serious injury, one would expect to find fewer hawks and more doves, since the average payoff to hawks will have declined. Similarly, if the value of the resource goes up, all things being equal, the proportion of hawks should go up while doves become increasingly scarce. This, too, makes sense; a more valuable prize warrants more risky behavior. And, finally, insofar as dove-dove confrontations might somehow become more expensive and time-consuming, this should give a push to hawkishness.

The key factor in such a system is the relationship between the value of the resource (food, in our case) and the cost of fighting over it. If the value of the resource is greater than the cost of trying to get it, then the hawk-dove game becomes a Prisoner's Dilemma, in which fighting (equivalent to "defect") is the dominant strategy, and hawks will take over . . . even though dove-dove cooperation provides a better joint outcome. But if the value of the resource is greater than the cost of trying to obtain it, then the Hawk-Dove Game is a Game of Chicken in which both hawks and doves can persist at a fixed proportion, determined by the exact values of the resource and its cost.

Aggressive displays between competing animals often boil down to Games of Chicken with a distinctly hawk-dove component. For example, two bull elk will begin their competition with a mutual roaring match, after which one individual may give up; that is, swerve, leaving the other in command of the field and possibly in possession of a harem of cow elk. If both bulls "defect," thereby carrying the confrontation to a higher and more dangerous level, there will be a stylized "parallel walk," after which the next stage involves strenuous pushing with locked antlers, and then, if need be, an all-out and potentially lethal fight. At each step, the contestants seek to overawe their opponent, inducing him to swerve in much the same way as participants in an automotive Game of Chicken.

Game theorists have focused much attention on iterated games of Prisoner's Dilemma. By contrast, Games of Chicken are generally considered to be one-time affairs. But in the realm of animal displays, with escalating levels of risk, we see fascinating examples of iterated Games of Chicken: hawk-dove interactions in which the players posture like hawks, each hoping to get the other to back down, while escalating to the next level if both remain steadfast.

Biologists have analyzed many variants on the Hawk-Dove Game. One is the so-called Game (or War) of Attrition. If there is a substantial cost to displaying, then "battles" of a sort could break out even among doves, since all things being equal it would pay the contestants to signal their intention of sticking it out, in order to try to induce the other to back down. After all, even though doves are peaceable creatures, more inclined to share a contested resource than to fight over it, each dove would nonetheless do better if it could get the whole thing, without getting violent (that is, without becoming a hawk). This is like signaling an intention to go straight in a Game of Chicken, for example by swaggering and swearing that you'll never swerve, come hell or high water. A big problem, however, is that the other player can't be confident that the boast is accurate; in fact, the signaler would be sorely tempted to lie, to claim it will endure at any cost, even though it won't, or can't. We often see this among our own species, when people play competitive risk-taking games. For example, in 1991, Iraq's Saddam Hussein swaggered about the forthcoming "mother of all battles." After his fall from power in Serbia, Slobodan Milosevic announced that he wouldn't be taken to jail alive. The Taliban forces in Afghanistan proclaimed that they, too, would fight to the last man, but didn't.

Although the Game of Attrition has its own special interest, the particular appeal of the Hawk-Dove Game is more fundamental. It provides an explanation for how two different behavioral strategies might coexist in the same population at the same time. After all, it is only a naïve view—of evolution, and of life itself—that suggests there should be only one best way of doing things. A game theory perspective, by contrast, concludes that whenever your payoff depends on what others are doing, it is possible for two contrasting strategies (such as hawks and doves) to coexist. This is especially true whenever increasing numbers of one strategy cause it to suffer a disadvantage, thereby tipping the scales in favor of the other.

Thanks to game theory and the hawk-dove model, biologists are now very much on the alert, their eyes peeled for situations in which two or more contrasting strategies sit comfortably on the same seesaw. Theoreticians, too, have gotten into the act, the result being a wonderful array of possibilities, some of which even seem to have parallels in the real world. For example, it is easy to see how hawks would profit in a world of doves, and also, how doves would do better than hawks in a population consisting largely of hawks. But at the same time, it is difficult to see how a very small number of doves, introduced into a population of hawks, could expand their numbers in the first place. When doves are rare, they not only do badly against hawks, but, in addition, they would only rarely encounter other doves. It is clear that, at this stage, doves need a boost. For one answer, game theoreticians have looked toward models in which doves interact preferentially with other doves, in a kind of benevolent conspiracy.[6]

The problem just described is that a world with hawks and other defectors would resist invasion by anyone inclined to be more cooperative, so long as these kindly new arrivals are all alone and cannot find like-minded individuals with whom to interact. But if the doves arrive in friendly, interactive cliques, they could happily exchange benefits with one another, and thereby prosper. In such a case, hawks will only rarely get to interact with doves (that is, those playing nasty will not have much chance to take advantage of those playing nice), and will have comparatively little opportunity to profit at the latter's expense. Most of the time, hawks would have to deal, like it or not, with others of their ilk, and suffer the consequences. By contrast, doves will do most of their interacting with other doves, to their benefit. Animal behavior and game theory thus explain, to some extent, the development of communities of like-minded individuals—even animal idealists!

Game theory guru Ken Binmore offers this psychologically realistic twist on the Hawk-Dove Game.[7] Imagine that both individuals have promised to play dove with the following payoffs, but only player 1 knows what the payoffs are:

		Player 2	
		Dove	Hawk
Player 1	Dove	2	0
	Hawk	3	1

This is not an unrealistic assumption; after all, people often find themselves promising to "be nice." If everyone plays dove, as promised, then everyone gets +2. But in this case player 1, despite what he or she promised, would be tempted to cheat, and play hawk instead, so as to get a payoff of +3. By doing so, player 1 would also ensure against being suckered into receiving payoff 0, in the event that player 2 cheated. But Binmore's twist introduces an element of self-punishment: Imagine further that if a player cheats, and acts the hawk, then he or she gets the predicted hawkish payoff, but also suffers some remorse, call it "r." Naturally, the greater the remorse, the greater player 1's payoff is diminished:

		Player 2	
		Dove	Hawk
Player 1	Dove	2	0
	Hawk	3-r	1-r

When r is greater than 1, interesting things happen. In particular, the payoff for being a hawk drops below that of being a dove, so dove replaces hawk. The moral: If social learning and cultural traditions make people feel sufficiently guilty about cheating, they won't do it.

Do animals feel guilt? Hard to say, but Waldeau, our family's golden retriever, certainly seemed to fit the bill. He had been told, trained, warned, and exhorted *not* to sleep on our down comforter, yet one day we came home to find Waldeau curled up in evident contrition, looking downright guilty. Sure enough, up on our bed there was a concavity in the comforter, lined with telltale wisps of golden hair!

Yet outward manifestations of guilt would probably be selected against in most circumstances, since it isn't in the interest of a guilty individual to advertise that fact. Instead, evolution seems more likely to favor individuals who hide their guilt, or—better yet—who don't feel any at all. This, in turn, might help explain the existence of sociopaths, transgressors who take advantage of others, all the while feeling no remorse, or sense of responsibility to others.

It is also possible that the animal equivalent of sociopathy is restrained if free-living creatures enforce a cost on transgressors, simply by punishing those who defect. For example, if a defector is subsequently ostracized because he or she behaved badly in the past, this would be precisely analogous to Binmore's "remorse"—and the greater the punishment, the greater the remorse.

For yet another twist, imagine a game in which each player empathizes with the other, feeling the other's pain but also sharing his or her gain. This can be modeled by giving each player a fraction, e (for empathy), of the other's payoff.

		Player 2	
		Dove	Hawk
Player 1	Dove	2 + (2 × e)	0 + (3 × e)
	Hawk	3 + (0 × e)	1 + (1 × e)

In this case, a dove encountering a hawk gains more than the usual 0, because the dove also receives a portion of the hawk's payoff. By the same token, a hawk encountering a dove gains nothing beyond a direct payoff of 3, since the dove's 0 payoff prevents the hawk from gaining any additional empathic glow. In this system, doves do better against doves than hawks do against hawks, since doves get an extra empathic payoff (2 × e) whereas hawks get only an additional 1 × e. It is worth paying special attention to what happens when the value of e exceeds one-half. Once again, doves enjoy a special payoff, because they gain an empathic benefit whether they are playing against a hawk or a dove, while hawks lose out on empathy if they take advantage of a dove.

In short, when empathy is afoot, niceness can be its own reward . . . but, sadly, empathy itself presumes a kind of underlying niceness. Maybe lack of empathy is one prerequisite for being a hawk!

The above game was set up so as to provide an extra payoff to doves, but, regrettably, there is nothing inherently favorable to doves in Hawk-Dove Games generally. It is equally possible to create a game in which hawks receive an extra benefit by taking pleasure in the *low* payoffs received by doves; call it the Nya-Nya or Neener-Neener Game. Try modeling it yourself.

Bullies, Bourgeois, and Other Complications

The Hawk-Dove Game can be made ever more realistic, more fun, and more complicated . . . and, thus, ever closer to life itself. One of the more prominent examples involves strategies for taking advantage of the benefit that hawks derive when they are paired with doves.

As presented thus far, hawks and doves have been stuck with their specified strategy: once a dove always a dove, and the same for hawks. Yet animals and people are often more flexible; they change their tactics based on how their opponents act.* Make way, accordingly, for a third strategy, called retaliator (R). When paired with a dove, a retaliator acts like a dove, but if retaliator's opponent escalates—that is, if it is a hawk—then retaliator plays hawk, too. This can be modeled as a 3 × 3 game, with hawk, dove, and retaliator making up the rows and columns, so that there are nine different payoff possibilities. It can be shown that under these conditions, the stable equilibrium situation, in biological jargon, the "evolutionarily stable strategy," is to be a retaliator, because retaliators get an advantage when paired with other retaliators: they behave as doves, thereby avoiding dangerous hawk-hawk escalation.

Significantly, many animals do, in fact, behave as retaliators, escalating only in response to an opponent's escalation. For example, a defeated rhesus monkey will tolerate the minor damage inflicted by the victor's incisors, but if the latter escalates and uses its canines, a vigorous retaliation usually follows; that is, a defeated animal plays dove, but if the victor goes too far, the retaliator within is revealed.

You may have noticed that the basic Hawk-Dove Game is essentially a Game of Chicken: substitute hawk for "go straight" and dove for "swerve." Thus, as in a standard Game of Chicken, the worst payoff in the Hawk-Dove Game comes from being a hawk—who fights, which is to say, refuses to swerve—when paired with another hawk. The hawk-hawk payoff is equivalent to the punishment of mutual defection; in other words, a car wreck. Similarly, the highest payoff comes from being a hawk paired with a dove, equivalent to the temptation to go straight when the opponent swerves. By the same token, medium payoffs accrue to being a dove with another dove, or—somewhat worse—a dove encountering a hawk. As predicted for the standard Game of Chicken, animal participants in real Hawk-Dove Games do, in fact, attempt to intimidate their opponents: puffing themselves

---

*In this context, a tactic refers to how an individual behaves in a particular encounter; a strategy refers to the decision rule that results in the chosen tactic.

up, roaring, baring their teeth, beating their chests, and so forth. On the other hand, when push comes to shove, they sometimes turn aside at the last moment!

Thus, Ken Binmore[8] talks about a souped-up version of Chicken, which he designates, agreeably enough, Chicken Soup. To create this concoction, Binmore adds a strategy called bully, who never fights, despite pretending that he will. If a bully meets a dove, he acts like a hawk; if he meets a hawk, the bully acts like a dove. If two bullies meet, they divide the pot, like two doves. The bully strategy can be highly successful. (Just visit any playground.)

Mathematically inclined analysts have also developed a 4 × 4 game, with hawk, dove, bully, and retaliator. Remember that retaliator starts off acting like a dove, but becomes instantly hawkish if the other guy shows any hawkish tendencies. So retaliator acts like a hawk to hawks, like a dove to doves, like a hawk to bully (whereupon bully retreats into dovishness). As a result, retaliator is typically the big winner. It probably isn't a coincidence that retaliator is very similar to the TIT-FOR-TAT strategy that Robert Axelrod found to be so successful in his computer tournament.

Bully and retaliator introduce an important element into the basic Hawk-Dove Game. Recall that an equilibrium of, say, 8:13 hawks and 5:13 doves could be achieved either by this particular fixed proportion of individuals, each of which always used the same strategy, or by giving every individual the flexibility of being either a hawk or a dove, in which case each one varied his or her behavior by the proportions indicated.

Let's look more closely at this option of changing behavior from situation to situation. You could change probabilistically, without reference to anything else, and, say, be a hawk 8/13 of the time, and dove, 5/13. But a smarter plan would be to make use of so-called "conditional strategies," which is really the essence of being a bully or a retaliator. Conditional strategies involve what logicians sometimes call "if, then" statements. Being a retaliator, for example, means "*if* you meet a dove, *then* act like a dove; *if* you meet a hawk, *then* act like a hawk." Being a bully means "*if* you meet a dove, *then* act like a hawk; *if* you meet a hawk, *then* act like a dove."

In the real world, conditional strategies are much of what life is about. If you are a male, act this way; if you are a female, act that way. If you are a parent, do this; if not, do that. If you are young, do this; if old, do that. And so forth. When it comes to conflict situations, of which the Hawk-Dove Game is a simplified model, it is obvious that conditional strategies loom large. If you are strong (or unprincipled), be hawkish; if weak (or perhaps, moral), be dovish. If you are especially hungry (or horny) be a pushy hawk; if not, be a more acquiescent dove. One simple way this strategy often arises in the natural world has been termed "assessor." It proclaims: "If you are larger than your opponent, play hawk; if smaller, play dove."

All well and good, but, in fact, we don't really need game theory to illuminate much of this, which can be explained, for example, by simple learning rules or the connection between hormonal state and behavior. Game theory becomes especially illuminating among animals— and, of course, humans—when behavior is *not* so obviously related to asymmetries between the players. And so, we come, logically enough, to what are called "uncorrelated asymmetries," differences in strategies that don't make sense, that involve abiding by rules simply *because they are the rules*. And as we shall see, it often pays to be law-abiding, even when the laws are arbitrary, and even when they aren't enforced.

Ethologists (scientists studying animal behavior) have long been puzzled and intrigued by the phenomenon of territoriality. In particular, what strikes ethologists as especially interesting is the fact that animals are so respectful of territorial ownership: When an intruder encounters a territory owner, the owner nearly always wins. In fact, "fights" between proprietor and interloper are usually not even worth the name; the trespasser simply gives up. One can easily imagine a territory owner saying, in effect, "Hey, this is my piece of turf," whereupon a visitor—instead of contesting the terrain—simply says, "Sorry, I'm outta here." This outcome is highly predictable, and nearly independent of need, physical capability, and past experience. It is, in short, an asymmetry that is uncorrelated with anything, except for prior ownership of the particular territory.

This curious respect that animals show for each other's territorial claims led John Maynard Smith to describe a strategy known as bourgeois, since it reflects traditional bourgeois morality, which respects ownership for its own sake, independent of any other claim to legitimacy or even physical ability to enforce one's will. Maynard Smith asks us to consider a hypothetical 3 × 3 game, in which the players are hawks, doves, and bourgeois. Bourgeois individuals—like bullies or retaliators—can be either hawk or dove, but, in this case, their "decisions" are based solely on whether they own the territory upon which the imagined confrontation takes place. They behave as hawks if they are the owners and as doves if they are intruding on someone else's turf.

Using the same payoff values as proposed earlier for the standard Hawk-Dove Game, Maynard Smith showed that bourgeois is the sole successful strategy. Start with hawk-dove, dove-dove, and hawk-hawk encounters: they are the same as described earlier. Bourgeois-dove and bourgeois-hawk encounters are also unremarkable; since bourgeois is equally likely to be resident or intruder in each situation, bourgeois plays hawk half the time and dove half the time. The crucial interaction is when bourgeois plays against another bourgeois: In such cases, hawk-hawk escalations never occur. This is because if one of the players is the resident, the other must be a trespasser. By the arbitrary bourgeois convention by which they "play," the resident then acts as a hawk and the intruder as a dove.

Thus, bourgeois gains an advantage over dove and hawk, as follows. Bourgeois fares better against a hawk than another hawk would, since fights result only half the time, while hawk-hawk conflicts always result in violence. In addition, bourgeois fares better against a dove than a fellow dove would, since half the time the bourgeois is an owner and, hence, a hawk. Bourgeois will therefore increase when rare, since it does better against a hawk than a hawk does against a hawk (don't forget that only half the time—when the bourgeois is a territorial resident—do bourgeois-hawk encounters escalate into hawk-hawk fights, whereas hawk-hawk encounters always do). Similarly a bourgeois does better against dove than the dove does against another dove, because half the time, the bourgeois is an owner and hence, a hawk (and hawks do well against doves).

At this point, it might help to look at the Hawk-Dove-Bourgeois Game as a matrix. To keep things manageable, I've left out the cost of displaying; otherwise, the Hawk-Dove payoffs repeat those presented earlier, when we examined the standard Hawk-Dove Game.

	Dove	Hawk	Bourgeois
Dove	½ food	0	¼ food
Hawk	all the food	½ food, minus cost of fighting	¾ food, minus ¼ cost of fighting
Bourgeois	¾ food	¼ food, minus cost of fighting	½ food

Once they become abundant, bourgeois strategists do pretty well against other bourgeois: no fighting, since two individuals cannot both own the same property. Bourgeois do better against themselves, in fact, than hawks do against bourgeois, since when hawk meets bourgeois, fights erupt half the time, assuming that a bourgeois is an owner as often as he or she is an intruder. And finally, bourgeois do better against other bourgeois than doves do against other bourgeois, because in the former situations, bourgeois get the benefit of being a hawk paired with a dove half the time (when bourgeois are owners), whereas in the latter, doves encounter bourgeois who act as hawks half the time, as a result of which they get nothing. So, when rare, bourgeois will spread, and, once abundant, they will stay abundant. In this regard, they are unlike either hawks or doves in a simple 2 × 2 game, in which—as we have seen—either type suffers once it gains in frequency.

The bottom line: It can pay to practice bourgeois morality, and adhere to a "meaningless" convention for its own sake.

These results are somewhat counter-intuitive, since they suggest that contests can be resolved by mutual adherence to a purely arbitrary rule, one that needn't bear any relationship to the player's abilities or needs. But, in fact, the benefits of abiding by such rules are clear to most people, and not surprisingly, human beings make use of them all the time. We often settle disputes, for example, by drawing straws, flipping a coin, or playing Paper-Rock-Scissors. As mentioned, ethologists have long recognized that territory owners generally win their disputes, although prior to the elaboration of game theory, they hadn't considered that the animals might be employing an "arbitrary asymmetry" such as the rules of prior ownership, uncorrelated with anything else.

Of course, there are confounding factors, not the least being that territory owners may have gained their property *because* they are somewhat bigger or fiercer. It shouldn't, therefore, be surprising if proprietors usually win, just as the majority of boxing matches between a champion and a challenger result in victory by the champ. (In such cases, we don't have to assume that the challenger is liable to lose because he is showing bourgeois respect for the champ's title, but, rather, because the champ became champ for good reason.) By the same token, if possession of a territory is crucial to survival or reproduction, and if there are limited numbers of suitable territories, bourgeois regard for arbitrary, uncorrelated factors would not be expected. The result, instead, is likely to be, if anything, a disregard for the rules— what evolutionary geneticist Alan Grafen calls the "desperado effect."[9] When times are tough, bourgeois morality may well go out the window.

Nonetheless, bourgeois morality abounds, and not just among *Homo sapiens*. Territory owners typically win, and not simply because they are bigger or stronger; in most cases, once an owner shows up, challengers give up without even appearing to try. The "home court advantage" is well known in human sports, and it seems to apply to birds as well. In one experiment, dark-eyed juncos were kept in groups of six until a clear social hierarchy was apparent. Then the top three from two separate groups were housed in the same cage. Territoriality prevailed: Whichever group of three was in its home cage assumed rank above the intruding group. Comparable results were obtained when the bottom ranking trios were brought together: birds 4, 5, and 6 became birds 1, 2, and 3 when combined, in their home cage, with another group of individuals who had ranked 4, 5, and 6 earlier.[10]

What about animals outside of cages, living in nature? An oft-cited research report on this question, titled "Territorial Defense in the Speckled Wood Butterfly" said it succinctly in its subtitle: "The Resident Always Wins."[11] At night, speckled wood butterflies sleep in trees, but as day advances, many of the males move to the forest floor, where each seeks to defend a small spot of sunlight as it moves about. These small territories are attractive to females, so males compete to defend them. If a female approaches a defending male and his illuminated patch of ground, she is courted; if a male approaches a defending male, a brief contest ensues in which the two butterflies spiral up in vertical flight.

Here is the key observation: In all observed encounters between resident and intruder, the resident *always* won, and without an escalated fight. Furthermore, when the researchers removed the resident, he was quickly replaced by another, who became the new proprietor. And when the former owner was then released, just a few minutes later, the *new* resident always won! Victory was achieved simply based on the fact of immediate, prior residence, not whether a particular competitor was larger, stronger, or fiercer. Finally, when the researchers surreptitiously introduced two males onto the same spot, it was possible to deceive each into thinking he was the resident. In such cases, the ensuing spiral flights lasted ten times longer than when one contestant "knew" that he was an intruder.

According to Nick Davies, who conducted this study, the typically brief spiral flight is a resident's way of announcing ownership, whereupon the intruder quickly acknowledges his error. The conclusion is unavoidable: It isn't really such a jungle out there. Or, at least, there are bourgeois butterflies in the bushes.

Following this initial research, which was conducted during a period of bright sunshine, other investigators studied the same species during cloudier weather. They found that in such situations, when sun-drenched territorial spots were harder to come by, intruders sometimes fought with residents, and that the previous owners usually won, but not always.[12] With the cherished resource of sunny spots in short supply, the payoff for fighting becomes greater, just as we saw in the standard Hawk-Dove Game.

Primates, too, practice middle-class morality. For example, male hamadryas baboons keep a harem and are rarely challenged by other males. Much as bachelor males might lust for their own contingent of females, they evidently recognize that a harem holder has certain rights of ownership. Thus, when a captive hamadryas male was allowed to associate with a female for just twenty minutes before a second male was released, the newcomer did not challenge the first male's "ownership" of the female. Such good manners were clearly not due to the fact that the first male was inherently more intimidating, since when the situation was reversed shortly afterward, the same male who had been deferred to now refrained from challenging the second.[13] (It seems that the leap from butterfly to baboon, in some ways, isn't all that far!) It is also interesting to note that when a free-living male hamadryas baboon is removed from his harem, his position is quickly taken by another. If the original harem holder is then released several weeks later, a vicious, escalated fight ensues, of the sort not normally seen; both males now behave as residents and use hawklike tactics, just like speckled wood butterflies when both are misled into thinking that each is the rightful owner of the same tiny territory.

It is strange to consider that animals evidently use arbitrary asymmetries to avoid potentially damaging hawk-hawk-style escalations. Strangest, perhaps, is that these asymmetries needn't accord with our usual (bourgeois) expectations. In theory, the arbitrary convention could just as well read, "Intruder always wins and resident always loses," and, indeed, there has been one report that a species of Mexican spider does just this: Residents give way to intruders, which produces a domino effect whereby a wave of resettlement sweeps through the local population once a single doughty individual enters another's residence.[14]

Arbitrary asymmetries may be part of the unacknowledged fabric of human society, where they function—as among hamadryas baboons and speckled wood butterflies—to minimize trouble. The best example may be one of the coordination games that we visited in chapter 2, resulting in the convention of driving only on one side of the road (the right side in the United States, the left in Great Britain). The origin of each rule is obscure, but it is nonetheless clear that once established— perhaps because a small number of intimidating drivers arbitrarily began one pattern or the other—any individuals who deviated would be penalized, not just legally, but by their own personal risk.[15]

What's arbitrary, therefore, is the method of settling disputes (whether by flipping a coin, playing a game of Paper-Rock-Scissors, deciding that "the resident always wins," and so forth). What isn't arbitrary is the underlying idea: the success of strategies that settle conflicts with a minimum of violence.

Dads, Cads, Coys, and Sluts

The natural world is usually analog rather than digital. Unlike the world of whole numbers, for example, which is neatly dichotomized into odd or even (but not in between), in the world of nature one thing slides into another, like day into night. But there are some important exceptions, one of the most prominent being male and female. To be sure, there are hermaphrodites—both male and female, simultaneously—as well as occasional individuals of indeterminate gender, but, by and large, the biological world is neatly cleaved into boys and girls, men and women, males and females. If you reproduce sexually, you are pretty much obliged to employ one strategy or another: either make a relatively small number of sex cells (eggs), each of which represents a large investment on your part, in which case you are a female, or make a very large number of sperm, each of which is tiny and trivial, in which case you are a male. (There is also an impressive body of theory and data showing that your optimum behavior is likely to be quite different depending on whether you are an egg maker or a sperm maker. But that's another story.[16])

You may have noticed that in discussing hawks and doves, we didn't concern ourselves with whether these characters were male or female. Let's change that, and take a quick look at sexual strategies. In this case, our question will be "How should you behave, given that you are one sex or the other?" But instead of examining conditional strategies (e.g., "If you are male and there are lots of other males around, be pushy, and if you are female . . ."), let's take a simpler case: namely, let's assume that there are two kinds of males and two kinds of females, and ask whether natural selection, abetted by game theory, will produce some kind of stable relationship among them. (The following discussion owes much to Richard Dawkins's innovative best-seller *The Selfish Gene.*)

Imagine that there are two male strategies, cads and dads, and two female strategies, call them coys and sluts. (As with the Hawk-Dove Game, there are many other possible strategies, but we'll stay simple.) Cads attempt to copulate with any females that come their way. They don't tolerate much delay for courtship, and after "having their way" with a female, they skip out, looking for other sexual opportunities. Cads are deadbeats in the dad department; their strategy is love 'em and leave 'em. Dads, by contrast, believe in courtship and also in sticking around and helping their mate rear offspring. Coy females are the egg-making equivalent of dads; as their name suggests, they are slow to warm up, and will copulate only with a male who spends time courting and making various demonstrations of his love, commitment, and fidelity. Sluts, on the other hand, are "fast," or "easy." They copulate readily, and without any discrimination.

It is reasonable to assume that there is a positive payoff to both parents if offspring are successfully reared, but also a cost in doing so. Kids are expensive. If this cost is divided equally between two parents, then both come out ahead, because the evolutionary payoff of reproducing is greater than the cost, so long as that cost is shared. But if one parent has to do all the work, then having kids is a losing proposition, at least in our hypothetical example. Let's also assume that there is a small cost to courtship, which takes time and energy, analogous to the cost of displaying when two doves meet in the Hawk-Dove Game. So, dads and coys are at a slight disadvantage to begin with (they court, unlike cads and sluts), but in the long run, they both stand to profit if they can find each other, since after investing a bit in delayed gratification, they get the benefit of each other's parenting efforts. At the same time, cads and sluts are ready to take advantage of anyone they can.

Start with a hypothetical population of devoted, responsible monogamists, inhabited by reliable males who are dads, and discriminating females who are coy. But now, introduce a rotten apple or two, equivalent to a hawk in a world of doves: Enter, a slut. She doesn't bother courting. Instead, she copulates promiscuously with the first male she finds. Why not? After all, there are only dads around, so she doesn't lose anything by having low standards, and she saves the cost of courtship. Our hypothetical slut makes babies with a dad, who—good dad that he is—struggles to rear the kids. As a result, slut genes get a higher payoff than coy genes, and sluts become more abundant, while coys start to decline.

But in a world of only coy females, males had to be dads, since any cads—who disdain courtship—were rejected and therefore didn't stand a chance. As sluts begin to prosper, however, there arises an opening for cads. After all, sluts don't discriminate between cads and dads, and thus cads have an advantage; since, like sluts, they don't waste time courting and, moreover, don't hang around being good fathers, either. Instead, they inseminate the sluts, then look for more victims. They are in pig heaven, getting a positive evolutionary payoff from all the babies they sire, without paying any of the costs borne by the more reliable dads. So, genes for cadishness start to spread, replacing those for dadishness.

At the same time, sluts are beginning to suffer. They still enjoy part of their old advantage over coys, in that they don't lose time courting, but now they find themselves increasingly abandoned by their caddy mates, forced to bear the large cost of single parenting, unless they elect to abandon their offspring and gain nothing. Coy females therefore start to get the upper hand, since even though they pay the small courtship cost, as before, they assure themselves the larger benefit of cooperative parenting from the dads they have chosen. We can also assume that when coy female meets cad male, no mating ensues (he wants, she refuses). She is waiting for Mr. Good-dad. At this point, the cycle of evolutionary balancing and rebalancing continues endlessly, since an increase in dads sets the stage for an increase in sluts—who, once again, don't waste time with courtship. And so forth.

The situation, modified here somewhat from John Maynard Smith's near-classic treatment,[17] is this:

If females are coy, it pays males to be dads.
If males are dads, it pays females to be sluts.
If females are sluts, it pays males to be cads.
If males are cads, it pays females to be coy. (Repeat indefinitely.)

Now, let's put some numbers to the theory. Imagine the genetic payoff to rearing a child is +15 for each parent, whereas the total cost of doing so is 20 for each child thus reared. So, if the expenses are shared equally between mother and father, then for every successful child, each parent gives up 10 (half the per-child cost of 20) and gains 15 for a net evolutionary profit of +5. Assume, also, that the cost of courtship

is 3. Time for another payoff matrix, considering the payoffs of producing one child:

		Female is	
		coy	slut
Male is	dad	2, 2	−8, 5
	cad	0, 0	15, −5

When dad male mates with coy female, and one child is reared, each
parent gains 15 while losing 10 in child care as well as 3 in courtship,
for a net payoff of +2. Our imaginary slut, introduced into this population, gains an additional +3 because she doesn't bother with courtship
(netting +5), and still gets a dad who rears the kid. So sluttishness
spreads. Until this point, cads couldn't get to first base, since all females
were coy. But as sluts become more abundant, things begin looking up
for the cads, who find willing partners and who don't pay any
courtship or rearing costs, but still get +15 for every kid. So they start
doing well, whereas the sluts begin suffering: Admittedly, they don't
pay anything for courtship, but they get socked with the entire −20 cost
of rearing each kid, who yields a profit of only +15. So, now that
they're mating with cads, every child costs them 5 fitness units.

Given these payoffs, there is an equilibrium that can be determined
mathematically; it occurs when 5/6 of the females are coy and 1/6 are
sluts, 5/8 of the males are dads and 3/8 are cads. At these particular
proportions, the average coy female does as well as the average slut,
and the average dad does as well as the average cad. So neither type
takes over, and neither goes extinct. Use different payoff numbers, and
you get different proportions, but, in any event, the stability is a fragile
one. If the proportion of the various types should happen to vary—by
chance—then the numbers can drift increasingly away from equilibrium. For our purposes, however, the important point is that the best
strategy, whether to behave like a dad or a cad, a coy or a slut, depends
on what others are doing. In our hypothetical world of sexual shenani-
gans, it is possible to reach an equilibrium point that is at least "stable
enough," in that different ways of behaving all coexist. Call it a tri-
umph of diversity.

Two mathematical biologists studying this system came to the following two-part conclusion: "(a) the battle of sexes has much in common with predation; and (b) the behavior of lovers is oscillating like the moon and unpredictable as the weather." And, they add, self-deprecatingly: "Of course, people didn't need differential equations to notice this before."[18]

Next, we turn to the real world of nature, and look at a few cases in which different lifestyles live together, if not amicably, then at least stably.

Different Ways of Being a Male

Supposedly, there is more than one way to skin a cat. (I don't know; I've never tried.) But in any event, it seems clear that there is more than one way to be a male animal. In this chapter, we'll therefore turn to "alternative male strategies." There has been very little research, incidentally, exploring whether the same applies to females: whether there is an array of "alternative female strategies." Maybe a future version of this book will describe, in delicious detail, how females are every bit as sneaky, weird, and diverse as males. But I doubt it. Sperm makers are under intense competitive pressure, since the payoff to males—lots of mates and, thus, lots of children—is generally high, just as the cost of failure is great: no children for the losers. By contrast, female animals can nearly always succeed simply by following the standard female strategy, whatever it happens to be. Not surprisingly, therefore, they seem less likely to indulge in strange alternative strategies.

In any event, we're going to explore the male situation, if only because that's where the evidence is.

For a beautifully analyzed case of alternative male strategies, we turn to insects known as scorpion flies.[19] There are three different ways for a male scorpion fly to acquire mates. He could be a tough guy, whose style is to drive other males away from a food source—generally a dead insect—to which females are attracted. Call it the Rambo tactic. Or he could be a spitter, who specializes in dribbling blobs of his own saliva onto leaves, then waits for females to come by and gobble up the tasty spitball. The third kind of scorpion fly doesn't offer anything to his potential mates; he simply grabs them and doesn't let go until the female copulates. Call him a rapist.

Biologist Randy Thornhill began investigating this system by putting ten male and ten female scorpion flies into cages. The males were of three sizes: small, medium, and large. Thornhill also placed two dead crickets in each cage. Not surprisingly, the largest males won the cricket competition, and these Rambos obtained the highest payoffs, averaging six copulations apiece. Medium-size males couldn't offer crickets to their lady loves, so they settled for being spitters, averaging a mere two copulations each. The smallest males were not only unable to defend a cricket, they also seemed incapable of making much spit. (Probably not scared spitless, their spitlessness was more likely metabolically imposed.) In any event, these small males became rapists, and their success was the lowest of all, only about one copulation apiece.

So far, these findings do *not* suggest that the Scorpion Fly Game of Rambo-Spitter-Rapist is like Paper-Rock-Scissors, or Hawk-Dove. This is because the three different "kinds" of male scorpion flies receive different payoffs, which means that in nature, Rambos should predominate; in fact, they should eventually drive spitters and rapists to extinction. So, why are there any spitters and rapists? The most likely reason is that male scorpion flies are not rigidly obliged to be either a Rambo, a spitter, or a rapist based on any fixed genetic program. (If so, then the only program that should become fixed is Rambo, the strategy whose payoff—and, thus, whose evolutionary fitness—is the highest.) Instead, the chances are that males are following a conditional strategy of this sort: If you can get away with it, be a Rambo; as a next-best option, be a spitter; if all else fails, be a rapist.

Thornhill reasoned that if this were the case, males should switch from a lower to a higher-paying strategy if given the opportunity. A simple test presented itself: Remove the Rambo males. An episode of musical chairs, scorpionfly-style, promptly ensued. Spitters abandoned their salivary secretions and became the new Rambos, lording it over the dead crickets that had previously been unavailable to them. And the former rapists—evidently still too small to put together a respectable gob of spit—stationed themselves alongside the nuptial gifts previously deposited by the spitters (now turned Rambos). Evidently, scorpion flies would just as soon move up if they could; in the meantime, they use a conditional strategy.

The Rambo-spitter-rapist triad seems unique to scorpion flies, but the general pattern of conditional reproductive strategies is widespread, especially among males, and even including mammals. For instance, some interesting complications arise among animals known as hoary marmots, relatives of the common woodchuck that inhabit the Cascade Mountains and northern Rockies. Male hoary marmots share their alpine meadows with one or two females, with whom they breed. But it's not quite that simple, since male marmots are also horny fellows, more than a little interested in breeding with neighboring females, who, in turn, are generally associated with their own males. Male marmots are therefore tempted to gallivant; that is, to wander about in search of receptive females. At the same time, however, a gallivanting male is vulnerable to gallivants by his neighbors, who are similarly motivated. It is possible, in fact, for a gallivanting male to achieve a tryst with a neighboring female, at the same time being cuckolded by the very male he is so busily deceiving!

But it doesn't happen often, simply because male hoary marmots follow a few basic strategies, which could have been prescribed by a game theoretician. They have two options: gallivant or guard. A moment's thought should convince you that if most of the surrounding males are gallivanting, a marmot concerned with his genetic posterity had better stay home and guard. On the other hand, if his neighbors are all guarding, there is little cost to gallivanting.

Admittedly, when most males are guarding, the average gallivanter will probably have little chance of success, since male marmots nearly always succeed in repelling intruders. (Despite their penchant for sexual exploits, they, too, are bourgeois territorialists.) Nonetheless, the sexual and reproductive payoff is potentially great, and the cost, small. So, males surrounded by guarders usually gallivant. And those surrounded by gallivanters usually guard.

During any given year, each female marmot is either fertile or not (each female reproduces every other year, although males are ready and eager to breed every spring). This leads to the following additional wrinkles: Male marmots don't gallivant when their females are fertile.

But when there are neither reproductive opportunities to be had at home nor reproductive risks incurred by playing the field, males are especially likely to gallivant.[20] However, as with the three types of scorpion flies, the two alternative strategies of male marmots are strictly of the "conditional" sort: "If this, then that." In addition—and once again, like scorpion flies—these different strategies are not fixed. The same male marmot will be a frisky sexual gallivanter under one set of circumstances, and a stodgy, stay-at-home mate guarder under others.

Multiple male strategies are found in a number of different species, typically separate, but rarely equal. On the other hand, there are also balanced combinations that reflect a genuine equilibrium among different kinds of "players," each of which are game-playing specialists, each sticking to a single strategy rather than being fair-weather conditionalists.

Perhaps the cleanest example comes from a species of isopod, relatives of the common sow bugs or pill bugs (those peculiar little armor-plated critters often found under rocks or logs). Our particular isopod lives in the ocean, specifically the Gulf of California. Wander about in the intertidal zone here, and you will find sponges. Open up the sponges, and you'll find female isopods, all of which are pretty much the same size. (As noted earlier, there aren't many good examples of alternative *female* strategies. At least, not yet.) But you'll also find three distinct kinds of males: big, medium, and small, with no intermediate forms. This in itself suggests that the individuals represent distinct genetic strategies. Otherwise, there should be intermediary forms as well: males who are small-to-medium, or medium-to-big, and so forth. And as it happens, the three different kinds of isopod males are distinct and discontinuous in their behavior, too.

Big males don't tolerate other males inside their sponge. If two big isopods encounter each other, they fight, sometimes nonstop for twenty-four hours, after which the winner takes possession and the loser leaves. If a big isopod gets his pincers on a small one, he unceremoniously tosses the little guy out. And so, small isopods avoid head-to-head meetings with big ones; instead, they lurk about, literally sponging off the biggies, grabbing the occasional opportunity to mate with an unguarded female. The medium isopods, meanwhile, are the same size as females, and they apparently can fool big males into thinking they *are* females! When a big male meets a medium one, the latter even acts like a female, soliciting courtship, and the big dummy obliges . . . but, of course, without any breeding as a result. In this way, medium isopods are allowed to hang around, becoming part of the retinue of deceived big males.

Studies have shown that these three different male isopods actually differ genetically,* giving further evidence that they represent distinct strategies maintained in evolutionary (and game theoretic) equilibrium. For this to be the case, however, each strategy would have to enjoy the same fitness; that is, equal reproductive payoffs. Researchers tested this by placing varying numbers of different kinds of males inside artificial sponges that were outfitted with appropriate numbers of females.[21] When cohabitors consisted of just one big male, one medium male, and one female, the big guy fathered most of the babies. But when comparable duos shared their spongy home with many females, medium males actually outcompeted the big dudes. (Not in head-to-head competition, but in the arena that really counts, biologically: reproducing.) Not only that, but when the female population was particularly large, sneaky and secretive small males actually had the most offspring of all! Even though big male isopods are more than a match for medium and small ones when paired mano a mano, they can't keep track of the female-mimicking mediums or the secretive smalls once things get crowded.

The final test of this system occurred when researchers sampled free-living sponges in the Gulf of California. They found that the three different types of isopod males had reproductive payoffs that were statistically indistinguishable.[22] In short, there appears to be nothing conditional about the marine isopod mating game. Unlike scorpionflies or marmots, big, medium, and small male isopods are separate, distinct, and coequal strategies, kept in balance by the fact that the payoff for each depends on what others are doing, each with its own niche, and with comparable evolutionary success.

---

*Technically, they each carry different forms—alternative alleles—of the same gene.

Only with the advent of game theory have biologists begun to look for alternative mating strategies of this sort. It has turned the old adage—"seeing is believing"—on its head: Believing is seeing. Now that biologists are believers in game theory, they've been seeing alternative strategies all over the place.

Take the case of bluegill sunfish, a common freshwater species not normally considered a game fish, although, as it turns out, its sex life is gamey indeed, both raunchy and remarkably game theoretic.[23] Among these fish, one can typically find "standard" big males: relatively large, often brightly colored macho fellows who aggressively defend a territory and "get" females the old-fashioned way. Then there are midsize males, "cross-dressers" who—like medium sponge-dwelling isopods—resemble females, and are thus able to fool the big guys, thereby getting mating opportunities by essentially hiding in the harem. And, finally, there may be some especially little fishies, often called "sneakers" or "streakers" who (similar to the small isopods) are typically smaller than females. In the case of bluegill sunfish, these streakers are aptly named, since they lurk on the outskirts of the action, only to swim suddenly amid a mating pair (a large sunfish male plus his inamorata), deposit a load of sperm, and zoom away. Paradoxically, these sneaky streakers are, in a sense, super-males, since they tend to have immense testes compared to their body size.

Finally, let's point out that males (and females, too, presumably) can coexist in ways that don't necessarily represent wholly alternative strategies. Consider two or more male lions cooperating—as they do—to maintain a pride of females. They typically do not fight with each other and, significantly, are often close relatives. Therefore, these males each receive a genetic payoff when the other reproduces. This is equivalent to the system of "empathic benefits" discussed earlier, except here the payoff is directly genetic. When players share genes, each receives a proxy benefit via the other's reproduction, with the exact amount of that benefit devalued in proportion as the two individuals are more distantly related.

John Maynard Smith has also noted that the close collaboration of male lions could be maintained by other factors in addition to kinship. In fact, unrelated males do cooperate on occasion. A single male, after all, would have virtually no chance of holding a pride of females against a united group of would-be usurpers; two males would stand a better chance; and three would be more likely yet to succeed (although the payoff to each would decline with increasing numbers, since the probability of fatherhood for any one cooperator would diminish).

In any event, mutual cooperation—even among nonrelatives—could be stable. Here is a possible matrix for the Harem-Share'em Game, with the payoff values indicating the number of offspring sired:

		Male B	
		cooperate	defect
Male A	cooperate	4, 4	0, 2
	defect	2, 0	1, 1

In this case, to cooperate means to share the effort and the benefit of working together to maintain and defend a harem. To defect means to insist on going it alone. If males A and B cooperate, they rear four offspring. This would be stable, since if male A defects, he does worse than if he stayed put (4 is better than 2), even though a defector would do better than male B who remains (defecting A would get 2, whereas cooperating B would get 0). Since the same applies to B, cooperation would seem assured. But this holds only if most lions are already cooperating—that is, if the system is already in the upper-left corner—so that any defectors are rare mutants.

In the matrix shown above, defecting can also be a stable outcome, an "evolutionarily stable strategy," if most individuals are doing it. To see this, start off in the lower-right corner, with everyone defecting and receiving a measly payoff of 1. Under this condition, a rare cooperator would receive 0, making it worse off than if it continued to defect. So, either mutual cooperation or mutual defection could be stable. What, then, leads to mutual cooperation, which is the usual situation? In the case of lions, it is probably kinship. This pattern could also be generalizable and may be responsible for much of the family-based cooperation found in other living things, including that peculiarly interesting creature, *Homo sapiens*.

7

Thoughts from the Underground

"Of all human pursuits," wrote St. Thomas Aquinas, "the pursuit of wisdom is the most perfect, the most sublime, the most profitable, the most delightful." Some renowned rationalists, from Voltaire to Tom Paine, found that their pursuit of reason placed them at odds with organized religion. Nonetheless, St. Thomas—no stranger to religion—suggested that "the proper operation of man is to understand." Before him, Aristotle maintained that happiness (not truth, or justice, or communion with God) is the ultimate goal of human life, arguing that happiness, in turn, comes from the use of *reason,* since that is the unique glory and power of humanity. Reason, for Aristotle—and to some extent, "the Greeks" generally—is the faculty that distinguishes us from all other living things, the life of reason being the greatest good to which human beings can aspire.

Nor were the Greeks alone in this belief. "What a piece of work is a man!" exults Hamlet. "How noble in reason! How infinite in faculty! In form and moving how express and admirable! In action how like an angel! In apprehension how like a god!" The Roman poet Virgil also spoke for many when he extolled the pleasure of using one's head: "Happy he who can know the causes of things."

All this gladdens the hearts (as well as the heads) of game theory devotees. After all, if game theory is about anything, it is reason:

more specifically, the application of rationality to certain kinds of decisions.

But what if rationality isn't all it's cracked up to be?

On Being Reasonable

When someone urges you to "be reasonable," the chances are that he wants you to see things *his* way. There must be a better definition.

According to the *Oxford English Dictionary,* "reason" is

> that intellectual power or faculty (usually regarded as characteristic of mankind, but sometimes also attributed in a certain degree to the lower animals), which is ordinarily employed in adapting thought or action to some end, the guiding principle of the human mind in the process of thinking.

This is worth pondering for a moment. Although we talk—and think—about reason as being somehow allied to sophisticated logical analysis, it may be best to define it more narrowly: as a means to an end. It is precisely in this sense that animals, as we have seen, often act "more reasonably" than do human beings, which enables game theory to offer unexpected windows into what makes us human.

We say that "there is no accounting for tastes," meaning that we cannot (reasonably?) claim that it is more rational to have one preference than another. It is not unreasonable to prefer ketchup over mustard. And although it may seem peculiar to like ketchup—instead of, say, chocolate sauce—on top of ice cream, reflection will probably suggest that such preferences cannot be adjudicated by competing claims of reason.

But—and here is the point—if you *do* prefer ketchup to mustard (or even to chocolate sauce), then it is irrational to buy mustard instead. And even more irrational to throw out your ketchup and use the mustard. (Unless you have a more important "reason" for doing so: Perhaps the ketchup is ridiculously expensive and you don't really prefer it to mustard by that much, so you purchase the mustard; or maybe your bottle of ketchup has gone bad, so you throw it away.) Rationality, in a sense, does not reside in one's preferences; rather, it is a tool for achieving these preferences, whatever they may be.

In the world of game theory, such thinking indicates that it is no more reasonable for a player to prefer one outcome over another. Given one's preferences, however, it may be more reasonable to proceed one way than another. Thus, it would be unreasonable to follow a strategy that leads to the alternative (less preferred) outcome, or to act in a manner that makes the preferred outcome less likely, or more delayed, and so forth. Reason, then, is a guide to achieving potential outcomes. It is a road map, but like any map, it can only tell us *how* to get someplace, not where to go. Let's face it: Although reason is an immensely powerful tool, it is also—like many power tools—difficult to use well.

One problem is just this: Reason isn't always easy. It is simple enough to figure out, for example, that "things equal to the same thing are equal to each other." If A equals B, and B equals C, then A equals C. But it is quite another to follow complex chains of tortured (but accurate) reasoning, no matter how "logical" the process.

Thus, sometimes we fall short of being rational not because we don't want to be, and not because of overpowering emotion, but simply because many of us aren't as good at it as we might wish. We fall prey to lots of logical fallacies. Here is a common one: Imagine that there are forty students in a classroom, and you are asked how likely it is that any two or more of them share the same birthday. The overwhelming majority of people respond, "*very* unlikely." After all, there are 365 days and 40 is only $\frac{1}{9}$ of this. "Common sense" suggests that perhaps only 1 time in 9 will there be two or more students with the same birthday. But in fact, with forty people, the actual odds are 7 to 1 *in favor* of this being the case.

Choose a student. The odds are 364/365 that a second student's birthday does *not* coincide with his. Now add a third. The chance that her birthday does *not* coincide with either of the first two is 363/365. As we proceed to all forty students, the chance that no two birthdays coincide is $364/365 \times 363/365 \times 362/365$ until we get down to 326/365, which multiplies out to approximately $\frac{1}{8}$. Remember, this is the chance that two of the students do *not* have the same birthday. Therefore, the chance that they *do* is ⅞. Another way of saying it: The odds are 7 to 1 in favor of such a coincidence. The key is that *any* two of your subjects can have the same birthday (which is much higher than the chance that if you choose any one individual, someone else in the class will have the same birthday). It is analogous to a roulette wheel with 365 slots, into which you dump forty different balls. What is the chance that *any* two will end up in *any* shared slot? Once again: ⅞. This is not the same as asking about the chance that two will end up on a predetermined date, say, April 23.

Reason can sometimes be terribly demanding, and we are not always up to the necessary mental gymnastics. Take this example, from Ken Binmore's textbook on game theory, cheerfully (if a bit misleadingly) titled *Fun and Games*.[1] Three people—Bob, Alice, and Nanny—are in a Victorian railway carriage, each minding his or her own business, reading a newspaper. Each has a dirty face. Let Professor Binmore tell the story from here:

> However, nobody is blushing with shame, even though any Victorian traveler who was conscious of appearing in public with a dirty face would surely do so. We can therefore deduce than none of the travelers knows that his or her own face is dirty, although each can clearly see the dirty faces of their companions.
>
> A railway guard now looks in through the window and announces that someone in the carriage has a dirty face. He is the sort of person who can be relied upon to make such an announcement if and only if it is true. One of the travelers now blushes. Why should this be? . . .
>
> To understand what is going on, it is necessary to examine the nuts and bolts of the chain of reasoning that leads to the conclusion that one of the travelers must blush. If neither Bob nor Nanny blushes, Alice would reason as follows:
>
> ALICE: Suppose that my face were clean. Then Bob would reason as follows:
>
> BOB: I see that Alice's face is clean. Suppose that my face were also clean. Then Nanny would reason as follows:

NANNY: I see that Alice and Bob's faces are clean. If my face were clean, nobody's face would be dirty, which I know to be false. Thus my face is dirty and I must blush.

BOB: Since Nanny has not blushed, my face is dirty. Thus I must blush.

ALICE: Since Bob has not blushed, my face is dirty. Thus I must blush.

Binmore uses this example to show the importance of knowing what others know: "Alice needs to know what Bob would know if she had a clean face. Moreover, she needs to know what Bob would know about what Nanny would know if Alice and Bob had clean faces. It is *this* information that the guard's announcement supplied." Indeed. My point is that in the real world, most people would not be capable of Alice's subtle reasoning, not to mention keeping up with Ken Binmore's explanation. They'd probably go on reading their paper, or would fall asleep and not bother to blush at all. Perhaps they should, however, not because reason makes an incontrovertible case that their face is dirty, but because they are probably unable to summon up the complexity of intellect needed to follow this chain of reasoning. (Give it a try: It works, but if you are at all like me, it will take a while, during which time your train may have arrived at its station.)

Even when the capacity exists, some people just aren't inclined to use reason terribly well, or especially often, or with the determination shown, for example, by Benjamin Franklin. In his *Autobiography*, Franklin described how he makes decisions:

My way is to divide half a sheet of paper by a line into two columns; writing over the one *Pro*, and over the other *Con*. Then, during three or four days' consideration, I put down under the different heads short hints of the different motives, that at different times occur to me for or against the measure. When I have thus got them all together in one view, I endeavor to estimate the respective weights . . . find at length where the balance lies . . . And though the weight of reasons cannot be taken with the precision of algebraic quantities, yet, when each is thus considered, separately and comparatively, and the whole matter lies before me, I think I can judge better, and am less liable to make a rash step; and in fact I have found great advantage for this kind of equation.

Many people do in fact follow at least some variant of Franklin's technique, listing *pros* and *cons* for certain decisions. But significantly, this is usually the case for quandaries that, in the great scheme of one's life, appear comparatively trivial: whether to buy a car, replace the computer, add another room to your house. When it comes to the really important things—whether to marry someone, or have children, or commit oneself to an all-encompassing political ideology or religious faith—rationality seems to be jettisoned altogether. Ironically, the more important the payoff matrix, the more difficult it is to fill it in, at least with anything approaching numerical accuracy. (What is the payoff of spending your life doing work that you find truly meaningful? Of living somewhere you can't stand? Of falling in love? As we learn from the musical *South Pacific*: "Who can explain it, who can tell you why? Fools give you reasons, wise men never try!") The really crucial payoffs just don't lend themselves to rational assessment. In short, when it comes to things that count, most of us don't bother counting.

On Being Unreasonable

Given that human thought is supposedly characterized by rationality above all else, it is more than a little paradoxical that so much impressive thinking has been devoted to debunking the role of rationality. We may speak admiringly of Greek rationality, of the Age of Reason, and of the Enlightenment, yet it is far easier to find great writing that extols unreason, irrationality, and the beauty of "following one's heart" rather than one's head.

For one thing, it may simply be that emotion is juicy, heart-thumping, and downright sexy whereas reason—by definition—is dry, head-pounding, often number-crunching, and only rarely making inroads below the waist. Persian poet Omar Khayyam made this trade-off uniquely explicit:

For a new Marriage I did make Carouse:
Divorced old barren Reason from my Bed
And took the Daughter of the Vine to Spouse

Thus presented, a rejection of reason seems, well, downright reasonable! Consider how rare it is for someone caught in the grip of strong emotion to be overcome by a fit of rationality, but how frequently events go the other way: After trying so hard to "be reasonable," we just "lose it," and succumb to our emotions, usually negative and troublesome ones at that. It is even possible to turn game theory on its head by emphasizing the role of emotions rather than reason in organizing our lives. In fact, maybe this is precisely what people—at least, some people—do, at least some of the time. When confronted with highly paradoxical situations, in which reason alone may be unhelpful, human beings can up the ante, or, in a sense, change the channel, by threatening, beseeching, yelling, crying, and so forth.

Ironically, discarding reason may be a kind of strategy—an unconscious one orchestrated by evolution—whereby people are essentially saying: "Better take me seriously on this, because I feel very strongly!" Consider how rarely, by contrast, we say, "I *think* this very strongly." Although people in the throes of an intense emotional reaction are unlikely to feel that they're being strategically savvy, seeking to manipulate the behavior of others by forcing them to reevaluate their own payoff matrix, this may be exactly what they are about. (Think—or feel?—back to those elephants in musth, in chapter 5.)

But no matter how fashionable it may be to "dis" reason, let's not be carried away.* Strong emotion can be wonderful, especially when it involves love. But it can also be horrible, when it calls forth hatred, fear, or violence. In any event, one doesn't have to idolize or idealize Greek-style rationality to recognize that excesses of unreason typically have little to recommend themselves, and much misery to answer for.

It is worth repeating that human beings—even the most rational

---

*By what? Presumably, by unreason, since as already suggested, people aren't generally swept away in an uncontrollable fit of rationality!

among us—are not strangers to irrationality. Legend has it that when Pythagoras came up with his famous theorem, justly renowned as the cornerstone of geometry (that most logical of mental pursuits) he immediately sacrificed a bull to Apollo! Or consider Isaac Newton: pioneering physicist, both theoretical and empirical, he of the laws of motion and gravity, inventor of calculus, and widely acknowledged as the greatest of all scientists. ("Nature and nature's laws lay hid in Night. God said, Let Newton be! And all was light.") This same Newton wrote literally thousands of pages, far more than all his physics and mathematics combined, attempting to explicate the prophecies contained in the Book of Daniel!

Go figure.

In Newton's case, as in Pythagoras', the most rarified rationality did not preclude unreason . . . or, as some would prefer to call it, faith. But at least, no great harm seems to have been done by the cohabitation. Sadly, this isn't always the case. "Only part of us is sane," wrote Rebecca West.[2]

> Only part of us loves pleasure and the longer day of happiness, wants to live to our nineties and die in peace, in a house that we build, that shall shelter those who come after us. The other half of us is nearly mad. It prefers the disagreeable to the agreeable, loves pain and its darker night despair, and wants to die in a catastrophe that will set life back to its beginnings and leave nothing of our house save its blackened foundations. Our bright natures fight in us with this yeasty darkness, and neither part is commonly quite victorious, for we are divided against ourselves. . . .

It may be significant that Ms. West wrote the above while reminiscing on her time spent in the Balkans, among inhabitants of what we now identify as the former Yugoslavia, people with a long, terrible history of doing things to one another that many outsiders readily label "insane" or, at least, "inhuman." Her point is deeper, however, not merely a meditation on Balkan irrationality, but on everyone's.

At this point, the careful reader might object: "Earlier, you wrote that reason does not dictate one preference over another, but only the best way to achieve such a preference. So how can you now agree with Rebecca West that it is irrational, or 'insane,' to prefer 'the disagreeable to the agreeable,' to love 'pain and its darker night despair,' to want to 'die in a catastrophe that will set life back to its beginnings and leave nothing of our house save its blackened foundations?' By your own argument, it is no less rational to prefer the one over the other!"

Although it is rationally appealing to define reason so as to exclude an individual's preferences, there is also a legitimate, commonsense insistence that certain choices—destruction over creation, death over life, chaos over order, unreason over reason—are, by definition, irrational. "Do I contradict myself?" as Walt Whitman asked. "Very well then I contradict myself. I am large, I contain multitudes." People contain multitudes; that's what makes them so fascinating.

Maybe it is "irrational" to accuse those who have destructive, pain-inducing preferences of being less than sane. But, if so, in the light of nuclear weapons and other devices of mass destruction, the world needs more such "irrationality." In some ways, it already has it. After all, to take an example from game theory, we have already seen how in single games of Prisoner's Dilemma, a rational player should always defect, but—fortunately—not everyone does. Cooperation can arise only if we can overcome our "rational" instinct toward selfishness, and, as we have been discussing, these instincts aren't always that strong. Which is just as well.

Similarly, it is clear that social dilemmas lead to situations in which individuals—if they were being rational—"should" behave in a way that is hurtful to the larger group. Often they do just this. But not always. Thus, people don't always follow the expectations of so-called "rational choice" models. Take voting. A famous analysis demonstrated with near certainty that the rational citizen should not bother to vote, because the time and effort required (even though minimal) could not be balanced by any conceivable positive payoff for the individual.[3] But people vote. Admittedly they do so in smaller numbers than many would like to see, but, nonetheless, it is striking that so many bother doing it at all.

In the case of voting, a fervent adherent of game theory and the role of rationality in motivating human behavior might well point to other considerations that play into the payoff matrix: socialization to be a "good citizen;" the warm glow of self-approval that comes from doing our "civic duty;" an egocentric, self-gratifying insistence that maybe—just maybe—our single vote really does matter; the pleasure of seeing our friends and neighbors at polling places. Put them together, and perhaps such "irrational" acts as voting turn out to be strictly "rational" after all.

More likely, I suspect, is the fact that when rational and irrational collide, people sometimes choose the irrational, to be cantankerous, for "reasons" that are genuinely unfathomable, or just "because." After all, Blaise Pascal, who abandoned his brilliant study of mathematics to pursue religious contemplation, famously noted "the heart has its reasons that reason does not know." Or as seventeenth-century English churchman and poet Henry Aldrich pointed out in his "Reasons for Drinking," sometimes we make up our minds first, and find "reasons" only later:

> If all be true that I do think
> There are five reasons we should drink:
> Good wine—a friend—or being dry—or lest we should be by and by—
> Or any other reason why!

Stubborn Unreasonableness

Of course, there are some who maintain that human irrationality and unreason are simply not comprehensible, and never will be. And furthermore, they argue, this is to be celebrated. Not for these people the hope expressed in Led Zeppelin's haunting song, "Stairway to Heaven":

> And it's whispered that soon
> If we all call the tune
> Then the piper will lead us to reason.

Would we all go marching, lock step and wholly gratified, following a Pied Piper of perfect rationality? Probably not. Not everyone admires reason. Religious mystics, for example, have long distrusted it. And excessive reason is easy to caricature.

Thus, at one point in Jonathan Swift's *Gulliver's Travels,* our hero journeys to Laputa, whose (male) inhabitants are utterly devoted to their intellects: One eye focuses inward and the other upon the stars. Neither looks straight ahead. The Laputans are so cerebral that they cannot hold a normal conversation; their minds wander off into sheer contemplation. They require servants who swat them with special instruments about the mouth and ears, reminding them to speak or listen as needed. Laputans concern themselves only with pure mathematics and equally pure music. Appropriately, they inhabit an island that floats, in ethereal indifference, above the ground. Laputan women, however, are unhappy and regularly cuckold their husbands, who do not notice. The prime minister's wife, for example, repeatedly runs away, preferring to live down on Earth with a drunk who beats her.

Michel de Montaigne devoted many of his essays to a skeptical denunciation of the human ability to know anything with certainty. But probably the most influential of reason's opponents was Jean-Jacques Rousseau, who claimed that "the man who thinks is a depraved animal," thereby speaking for what came to be known as the Romantic movement. Even before the advent of Romanticism, however, many thinkers, including those who employed reason with exquisite precision, had been inclined to put it "in its place." Hard-headed empiricist philosopher David Hume, for example, proclaimed that "reason is, and ought to be, the slave of the passions, and can never pretend to any other office than to serve and obey them."[4] Furthermore, when reason turns against the deeper needs of people, Hume argued, people will turn against reason. In any event, the most pronounced turning against reason occurred among the Romanticists, who embodied a rejection of what had seemed a sterile, dry, deadening exaltation of rationality as *the* distinguishing human trait. According to historians Will and Ariel Durant,[5] the Romantic movement constituted

> the rebellion . . . of sentiment against judgment, of the subject against the object, of subjectivism against objectivity, of solitude against society, of imagination against reality, of myth and legend against history, of religion against science, of mysticism against ritual, of poetry and poetic prose against prose and prosaic poetry, of neo-gothic against

neoclassical art, of the feminine against the masculine, of romantic love against the marriage of convenience, of "Nature" and the "natural" against civilization and artifice, of emotional expression against conventional restraints, of individual freedom against social order, of youth against authority, of democracy against aristocracy, of man versus the state.

But, most important, Romanticism embodied the rebellion of feeling against reason, of instinct against intellect. It isn't found in game theory but is reflected in the music of Mendelssohn, Schumann, Wagner, Chopin, and—the first great musical Romantic—Beethoven. Also the painting of Delacroix, Constable, and Turner, as well as the poetry of Coleridge, Keats, Shelley, Byron, and especially Wordsworth. More recently, modernism and postmodernism can be seen as yet another revolt against the seemingly sterile—and often downright destructive—consequences of reason, especially in the form of mass society and increasingly lethal technology. (After all, the gas chambers of Auschwitz and the nuclear devastation of Hiroshima and Nagasaki were triumphs of human reason; if they are to be criticized, it must be on some other basis than as failures of rational design.)

Probably the most articulate, not to mention downright angry, denunciation of human reason is found in the work of Fyodor Dostoyevsky. In his novels—notably *Crime and Punishment, The Brothers Karamazov, The Possessed,* and *The Idiot*—Dostoyevsky explored the darker, irrational side of human nature. Dostoyevsky's dissatisfaction with rationality is consistent with his stance as a "Slavophile," part of a centuries-old schism in Russian thinking between the "Westernizers," who looked to Europe (and hence, to science, reason, and "modernity") and the Slavophiles, who favored a Romantic, near-mystical belief in Russian greatness as well as in the power of history, religion, irrationality, the peasantry, and the land. (It is a dichotomy that can still be identified in Russia today.)

Dostoyevsky's most powerful—and uniquely disturbing—critique of reason came in his short novel *Notes from Underground,* which depicted a nameless antihero: unattractive, unappealing, and irrational (although intelligent!). Recall the "utilitarians" such as Jeremy Bentham and John Stuart Mill, with their assumption that society should aim for the "greatest good for the greatest number," and that people can be expected to act in their own best interest. In recent decades, that same perspective has been the foundation of much economic theory (so-called "rational choice" models in which people are thought to maximize their "utility," or benefits from a given act) as well as evolutionary biology (in which living things are seen as maximizing their "fitness"). And, of course, game theory as well.

Dostoyevsky would have none of it. In *Notes from Underground,* the Underground Man—literature's first "antihero"—recounts various willfully painful and irrational experiences he has undergone, as he jeeringly argues that humanity can never be encompassed within a "Crystal Palace" of rationality. He may have a point: Certainly, unreason can be every bit as "human" as the Greeks believed rationality to be. You don't have to be a Freudian, for example, to recognize the importance of the unconscious, which, like an iceberg, not only floats largely below the surface—and is thus inaccessible to the cold light of reason—but also constitutes perhaps 90 percent of our total mental mass.

It is one thing to acknowledge the importance of unreason and irrationality, and quite another to *applaud* it, as the Underground Man does: "I am a sick man. . . . I am a spiteful man. I am a most unpleasant man." The key concept for Dostoyevsky's irrational actor is *spite,* a malicious desire to hurt another without any compensating gain for the perpetrator. Consider the classic formulation of spite: "cutting off your nose to spite your face," disfiguring yourself for "no reason." Causing suffering, embarrassment, or pain, and why? No reason! Just spite.

Significantly, spiteful behavior does not occur among animals. Even when an animal injures itself or appears to behave irrationally—gnawing off its own paw, killing and eating its offspring—there is typically a biological payoff: freeing oneself from a trap, turning a child (who under certain circumstances may be unlikely to survive) into calories for the parent. Spite is uniquely human.

The Underground Man goes on to rail against a world in which—to his great annoyance—two times two equals four; claiming that there is pleasure to be found in a toothache; referring, with something close to admiration, to Cleopatra's alleged fondness for sticking golden pins in her slave girls' breasts in order to "take pleasure in their screams and writhing."

As the Underground Man sees it, the essence of humanness is living "according to our own stupid will." It must embrace the right—even, the obligation—to "desire something even very stupid and not be bound by an obligation to desire only what's smart . . . because it preserves for us what's most important and precious, that is, our personality and our individuality."

We've already seen that much of animal behavior is, paradoxically, closer to the rational dictates of game theory than are the actions of human beings. But this appears to be due to the fact that animals are more closely manipulated by their genes, and these genes, in turn, have been sifted and winnowed by natural selection, leaving as "winners" those that maximize their genetic representation by corralling the largest possible payoffs. Insofar as this is true, then human behavior is less precisely calibrated than that of animals, but not for reasons that the Underground Man would approve. He believes that people act irrationally because they stubbornly *want* to do so, not because they aren't "smart enough" or "evolved enough" to know, automatically, their best interest. In fact, the whole concept of "best interest" (substitute, if you wish, "maximizing one's game theoretic payoff") is anathema to him:

> Shower him with all sorts of earthly blessings, submerge him in happiness over his head . . . even then, out of pure ingratitude, sheer perversity, he'll commit some repulsive act. . . . If he lacks the means, he'll cause destruction and chaos, he'll devise all kinds of suffering and have his own way! He'll leash a curse upon the world; and, since man alone can do so (it's his privilege and the thing that most distinguished him from other animals), perhaps only through this curse will he achieve his goal, that is, become really convinced that he's a man and not a piano key!

And then, as though speaking directly to a game theoretician, the Underground Man snarls, "If you say that one can also calculate all this according to a table, this chaos and darkness, these curses, so that the mere possibility of calculating it all in advance would stop everything and that reason alone would prevail—in that case man would be insane deliberately in order not to have reason, but to have his own way!"

Such sentiments are in no way limited to the most famous apostle of the dark Russian soul, or to nineteenth-century European Romantics. Here is a poem from that quintessentially American writer Stephen Crane:[6]

> In the desert
> I saw a creature, naked, bestial,
> Who, squatting upon the ground,
> Held his heart in his hands,
> And ate of it.
> I said, "Is it good, friend?"
> "It is bitter—bitter," he answered;
> "But I like it
> Because it is bitter,
> And because it is my heart."

Predictable Unreasonableness

The Underground Man prides himself on being unreasonable, and thus, unpredictable. Scientists prefer to study patterns that are predictable, even if not strictly "reasonable."

In a now-classic experiment, psychologist L. G. Humphreys sat his subjects down in front of two lightbulbs. Their task was to predict which of the two bulbs would light up next. In fact, they went on and off according to a random schedule, with one of them (say, the right) going on 80 percent of the time and the other, 20 percent. Initially, Humphreys's subjects guessed that the two bulbs would light up equally often. (This makes a kind of sense, although, in fact, any prediction would have been equally "logical.") Over time, however, the right one continued to blink on and off four times more often than the bulb on the left, and sure enough, the subjects began to favor that one. After much experience with the system, when all the subjects had become very familiar with the pattern, nearly all of them followed the same strategy, one that was suboptimal and, thus, irrational: They distributed their guesses according to the proportion with which the bulbs actually lit up. That is, they guessed "right" 80 percent of the time and "left" 20 percent.[7]

This is *not* the best strategy, although it initially seems logical. The best strategy—the one most likely to result in correct responses—is to guess that the most frequently flashing bulb will be the one illuminated *every time,* since even if you know the overall probabilities, you still don't know what will happen on any given event. In Humphreys's experiment, someone who followed this strategy and constantly picked the bulb on the "right" would be correct 80 percent of the time. By contrast, as a result of guessing "right" 80 percent of the time, a "matcher" would be correct 80 percent × 80 percent = 64 percent (since 80 percent of the time the bulb will light, and 80 percent of the time, the "matcher's" guess will correspond), plus 20 percent × 20 percent = 4 percent (20 percent of the time the left bulb will light, and the "matcher" guesses left 20 percent of the time) for a total of only 68 percent correct.

As to why people behave so consistently, and yet so wrongly, there are many ideas. Perhaps the most likely also explains why investors are often inclined to try selecting individual stocks, or—at minimum—to choose a mutual fund led by someone they believe to be an especially good stock picker, even though index funds outperform either strategy more than 90 percent of the time: People like to think that they are special, and, thus, better or luckier than the average. We might call it the Lake Woebegone syndrome, after Garrison Keillor's fictional community where "all the children are above average." Thus, investors resent being at the mercy of larger forces, told that the best they can get is what everyone else will obtain. By the same token, maybe lightbulb predictors can't help thinking that once they are aware that the left bulb occasionally lights up, maybe they'll be able to anticipate when. Placing all one's bets on the right-side bulb feels too unresponsive to the widespread sense that each of us is uniquely special and competent, too much like buying an index fund.

Psychologists have been intrigued with consistent departures from rationality of this sort, and have cataloged numerous examples,[8] including the following:

> Imagine that you have decided to see a play and paid the admission price of $10 per ticket. As you enter the theater, you discover that you have lost the ticket. The seat was not marked, and the ticket cannot be recovered. Would you pay $10 for another ticket?

Forty-six percent of experimental subjects answered yes; 54 percent answered no. Then, a different question was asked:

> Imagine that you have decided to see a play where admission is $10 per ticket. As you enter the theater, you discover that you have lost a $10 bill. Would you still pay $10 for a ticket for the play?

The results: This time, a whopping 88 percent answered yes and only 12 percent answered no. In other words, most people say that if they had lost their ticket, they would be unwilling to buy another, but if they had simply lost *the value* of the ticket ($10), most have no qualms about making the purchase! Why such a huge difference?

It is irrational, but perhaps explicable, according to psychologists Daniel Kahneman and Amos Tversky, by the way people organize their mental accounts:

> Going to the theater is normally viewed as a transaction in which the cost of the ticket is exchanged for the experience of seeing the play. Buying a second ticket increases the cost of seeing the play to a level that many respondents apparently find unacceptable. In contrast, the loss of the cash is not posted to the account of the play, and it affects the purchase of the ticket only by making the individual feel slightly less affluent.[9]

Here is another one: "Would you accept a gamble that offers a 10 percent chance to win $95 and a 90 percent chance to lose $5?" The great majority of people in the study rejected this proposition as a loser. Yet, a bit later, the same individuals were asked this question: "Would you pay $5 to participate in a lottery that offers a 10 percent chance to win $100 and a 90 percent chance to win nothing?" A large proportion of those who refused the first option accepted the second. But the options offer identical outcomes. As Kahneman and Tversky see it: "Thinking of the $5 as a payment makes the venture more acceptable than thinking of the same amount as a loss." It's all a matter of how the situation is framed.

To some degree, these cases don't necessarily depict irrationality, so much as the fact that rationality itself isn't always obvious, or agreed upon. For example, it is well established that many more people would accept a sure $100 rather than take an equal chance on winning $200 or nothing, although in this case, the payoffs are mathematically identical. People are what psychologists call "risk averse," which isn't necessarily irrational, but isn't strictly rational either.

Psychologists and others who work in "decision theory" are also aware of what is known as "Allais's paradox." Economists had long assumed that people assess probabilities in proportion as they are, in fact, probable; that is, a 90 percent probability is known to be 1 percent more likely than an 89 percent probability, and so on, from probabilities of 0 to 100 percent. But no. Many experiments have shown that the difference between 99 percent and 100 percent has a great impact on people's decisions, far more than, say, increasing a probability from 29 percent to 30 percent. Similarly, going from 0 to 1 percent has more impact than going from 12 percent to 13 percent. (This is an example of "nonlinearity of preferences": An event has more psychological impact if it changes a possibility into a certainty—99 percent to 100 percent—or impossibility into possibility—0 to 1 percent—rather than simply increasing a possibility. This holds even if the "actual impact" of each of these changes is logically identical.)

Then there is something called "source dependence": preferring, for example, to bet on a proposition drawn from one source than from another, even if the chances of winning are the same. A good example is the "Ellsberg jar." Imagine a jar containing one hundred balls, fifty red and fifty black. Another jar also has a hundred red and black balls, but you don't know the proportion of either color. Now, imagine further that you will be given $50 if you can guess the color of the ball drawn from one of these jars. Which jar would you choose? Most people would rather choose jar number one: better the probability you know than the one you don't.[10]

A last example:

> Assume you have been exposed to a disease which if contracted leads to a quick and painless death within a week. The probability you have the disease is 0.001. What is the maximum you would be willing to pay for a cure?

A typical response in this case is $200. Now think about this question:

> Suppose volunteers were needed for research on the above disease. All that would be required is that you expose yourself to a 0.001 chance of contracting the disease. What is the minimum payment you would require to volunteer for this program?

Here, a typical response is on the order of $10,000, which is fifty times higher than in the earlier situation. And yet, the two are logically identical! It is much harder to get people to volunteer—to initiate an action—than to deal with something that may have already happened to them.[11]

Examining these and other cases of human irrationality, or errors in decision making, psychologists have been struck by their "robustness," that is, the degree to which they are repeatable. In fact, these patterns are so robust that various generalizations have been proposed: "risk aversion," "framing effects," "prospect theory," and so forth.* Presumably, these patterns will eventually become explicable and not just nameable, as our understanding of human behavior—including nonrationality—becomes more sophisticated. Maybe someday the various manifestations of human nonrationality will actually be *understood*.

---

*In 2002, Daniel Kahneman—like John Nash before him—won a Nobel Prize in economics, this time for showing how human psychology, with all its nonrationality, influences decision making.

Understandable Unreasonableness

It is probably just as well that the Underground Man is not in the majority, or, at least, not obviously so. We may shake our heads, knowingly, at his insistence on being unpredictable, even unpleasant, spiteful, or downright stupid. But most of us wouldn't choose him to be our financial adviser, or vocational adviser, or romantic adviser, or, indeed, any sort of purveyor of wisdom. Maybe reason doesn't make the heart sing, but as a guide to action, it is probably a lot better than its darker, danker, likely more destructive, albeit sexier alternative.

Reason may not always carry the day, but it is more likely to elevate than debase us. As Shakespeare noted in *A Midsummer Night's Dream,* "The will of man is by his reason swayed; And reason says you are the worthier made."[12] (Admittedly, reason is not a disinterested judge in this case. Should reason therefore recuse itself, and let raw emotion decide which is better?) In any event, and regardless of the merits of reason, there seems undoubted merit in trying to understand what sets the limits to rationality.

One possibility—as the Underground Man insists—is that people just don't want to be reasonable. Another, suggested in our earlier example of dirty-faced Victorians, is that they simply aren't up to the task. Nobel Prize–winning psychologist and early computer theoretician Herbert Simon gestured toward explaining these limits when he noted that human beings are restricted to "bounded rationality." By this he meant that

> the capacity of the human mind for formulating and solving complex problems is very small compared with the size of the problems whose solution is required for objectively rational behavior in the real world— or even for a reasonable approximation to such objective rationality.[13]

Game theory assumes that people seek to maximize their positive outcomes. We have already considered that some—such as Dostoyevsky's Underground Man—deny that this is even a goal. Others, such as Simon, believe (or is it "are rationally convinced"?) that logic-based maximization of payoffs validly describes much human striving, but that it is all too rarely achieved in practice. Simon emphasized, for example, that because their rationality is bounded by practical constraints, when it comes to evaluating outcomes in real-life situations, people are more likely to pay attention to the basic distinction "satisfactory/unsatisfactory" rather than to strive for perfect maximization, which often requires detailed attention to fine gradations. Accordingly, he introduced the term *satisficing*, as a hybrid of *satisfying* and *sufficing*.

Thus, not-so-simple Simon says that instead of obtaining perfection, people are likely to sample their options and choose one that is satisfactory because it suffices. This way, the searcher avoids spending too much time and energy, and also avoids encountering options whose complexity may be beyond his or her ken. Chess players, for example, cannot possibly evaluate all possible options: There are approximately thirty legal chess moves at any one time on a chessboard, and a game typically involves about forty moves by each player. Multiply 30 by itself, 40 times: that is, 30 to the 40th power, and you get 10 to the 120th power, for the number of possible chess positions in an average game. It is estimated that there are between 10 to the 78th and 10 to the 80th particles in the universe! So, a good chess player must satisfice. People looking for a new car, or a house, or even a mate, nearly always stop looking after finding something or someone who meets certain minimal, realistic requirements. Sounds hard-hearted, but it is probably true. Indeed, maybe "love" is that feeling of excitement and delight that comes from recognizing that someone else satisfices!

Which brings us to yet another perspective on why *Homo sapiens* aren't always as sapient as they might like, or as game theory assumes. We can agree with Herbert Simon that the mind is indeed incapable of solving many of the problems posed by the real world, not simply because the world is big and the mind is small, but bcause the mind did not develop as a calculator designed to solve logical problems. Rather, it evolved for a very limited purpose, one no different from that of the heart, lungs, or kidneys; that is, the job of the brain is simply to enhance the reproductive success of the body within which it resides. (And in the process, to promote the success of the genes that produced the body, brain and all.)

This is the biological purpose of every mind, human as well as animal, and, moreover, it is its *only* purpose. The purpose of the heart is to pump blood, of the lungs to exchange oxygen and carbon dioxide, while the kidneys' work is the elimination of toxic chemicals. The brain's purpose is to direct our internal organs and our external behavior in a way that maximizes our evolutionary success. That's it. Given this, it is remarkable that the human mind is good at solving any problems whatsoever, beyond "Whom should I mate with?," "What is that guy up to?," "How can I help my kid?," "Where are the antelopes hanging out at this time of year?" There is nothing in the biological specifications for brain building that calls for a device capable of high-powered reasoning, or of solving abstract problems, or even providing an accurate picture of the "outside" world, beyond what is needed to enable its possessors to thrive and reproduce. Put these requirements together, on the other hand, and it appears that the result turns out to be a pretty good (that is, rational) calculating device.

In short, the evolutionary design features of the human brain may well hold the key to our penchant for logic as well as illogic. Following is a particularly revealing example, known as the Wason Test.

Imagine that you are confronted with four cards. Each has a letter of the alphabet on one side and a number on the other. You are also told this rule: If there is a vowel on one side, there must be an even number on the other. Your job is to determine which (if any) of the cards must be turned over in order to determine whether the rule is being followed. However, you must only turn over those cards that *require* turning over. Let's say that the four cards are as follows:

T   6   E   9

Which ones should you turn over? (Remember, you want to assess this rule: If there is a vowel on one side, there must be an even number on the other.)

Most people realize that they don't have to inspect the other side of card "T." However, a large proportion respond that the "6" should be inspected. They are wrong: The rule says that if one side is a vowel, the other must be an even number, but nothing about whether an even number must be accompanied by a vowel. (The side opposite a "6" could be a vowel or a consonant; either way, the rule is not violated.) Most people also agree that the "E" must be turned over, since if the other side is not an even number, the rule would be violated. But many people do not realize that the "9" must also be inspected: if its flip side is a vowel, then the rule is violated. So, the correct answer to the above Wason Test is that "T" and "6" should not be turned over, but "E" and "9" should be. Fewer than 20 percent of respondents get it right.

Next, consider this puzzle. You are a bartender at a nightclub where the legal drinking age is twenty-one. Your job is to make sure that this rule is followed: People under twenty-one must not be drinking alcohol. Toward that end, you can ask individuals their age, or check what they are drinking, but you are required not to be any more intrusive than is absolutely necessary. You are confronted with four different situations, as shown below. In which case (if any) should you ask a patron his or her age, or find out what beverage is being consumed?

#1	#2	#3	#4
Drinking Water	Over 21	Drinking Beer	Under 21

Nearly everyone finds this problem easy. You needn't check the age of person 1, the water drinker. Similarly, there is no reason to examine the beverage of person 2, who is over twenty-one. But obviously, you had better check the age of person 3, who is drinking beer, just as you need to check the beverage of person 4, who is underage. The point is that this problem set—which is nearly always answered correctly—is logically identical to the earlier set, the one that causes considerable head scratching, not to mention incorrect answers.

Why is the second problem set so easy, and the first so difficult? This question has been intensively researched by the husband-and-wife team of Leda Cosmides, an evolutionary psychologist, and John Tooby, an evolutionary anthropologist (the recurrence of the word *evolutionary* isn't coincidental, as we shall see). Their answer is that the key isn't logic itself—after all, the two problems are logically identical—but how they are positioned in a world of social and biological reality. Thus, whereas the first is a matter of pure logic, disconnected from the real world, the second plays into issues of truth telling and the detection of social cheaters. The human mind, Cosmides points out, is not adapted to solve rarified problems of logic, but is quite refined and powerful when it comes to dealing with matters of cheating and deception.[14]

The card example just described is a special case of the more general phenomenon, known as a "conditional rule." Such rules are simple statements, familiar to logicians: "If P then Q." The rule is violated if, in a given situation, P is true and, at the same time, Q is false. There are lots of ways of stating such conditional rules, all of which are more or less familiar: for example, if someone lives in Seattle, he or she is standing in the rain; if someone loves America, then he or she voted for George W. Bush; if someone is skiing, then it is winter. It turns out that only about 25 percent of subjects do a good job of detecting violations of such rules. On the other hand, if the rule is something like this—"If you attend a movie, you must buy a ticket"—then people are much better at detecting violations: about 75 percent get it right!

In short, our rationality is indeed bounded, and not merely by our inherent limitations as small creatures in a large universe. More important, perhaps, we are bounded by what our brains were constructed—that is, evolved—to do. This does not necessarily bode ill for the human ability to behave in accord with the predictions of game theory, or to understand its ideas, since, as we have already seen, many of these games are truly games of life itself. But it seems very likely that in all cases, it is "life" that comes first, and games—whether logical or not—only later.

A Troublesome Fallacy

Some logical fallacies make a kind of sense, in that we can understand where they come from, and, thus, how they have arisen. Nonetheless, they remain deep-seated in the human psyche, and troublesome in their consequences. One of these is known as the "gambler's fallacy," or sometimes, "the Concorde fallacy." It resonates, as well, in the tragedy of the Vietnam War and in the fiendishly diabolical Dollar Auction Game devised by game theorist Martin Shubik.[15]

Let's start with the general fallacy. It is simple enough: Many people assume that if they have already committed time or money toward some end, they are therefore "logically" obliged to commit even more. "In for a penny, in for a pound," as the old English proverb has it. Gamblers often feel that once they have placed a bet, they have no choice but to continue betting, to stay "in" even as the ante is upped.

This is especially true among poker players. Let's say that on the first round of betting, they "see" their opponent; that is, they put up a certain amount of money to remain in the game. By the next round of betting, they may feel significantly less confidence in their cards, but, having already committed some money, they find it difficult to back out. And by the round after that, if there is one, they find it more difficult yet to "fold."

Biologists Richard Dawkins and Tamsin Carlisle suggested calling it the Concorde fallacy, after the insistence by the British and French governments that they must continue to spend money developing the Concorde Supersonic Transport plane, even after it was demonstrated that the project would not be economically feasible. "We've already spent several hundred million pounds (francs)," it was argued, "so we can't pull out now." To back out, after investing so much money, was "illogical," since it would mean that the expenditure was in vain.[16]

But the only logical strategy—for gamblers or governments—is to assess the best possible "move" at any given time. It doesn't matter how much, or how little, has already been spent. What's gone is gone. The question at any given time is: "What is the most effective way to invest right now, regardless of how much has already been committed to any particular project?" But the human mind resists such logic.

Why? Probably because even though the quantity of past investment does not logically have to correlate with the wisest current expenditure, it is likely that for many biological situations, it has done so. If you have already spent a lot of time and energy rearing a child, the chances are that this child is older, larger, wiser, and more healthy than another child with whom you haven't been comparably involved. Similarly, if you have already spent a lot of time and energy building a nest, the chances are that it is near completion, so—like the hypothetical child—it probably won't require much more effort to finish the job. In short, there may be a kind of biologically savvy rule of thumb that says, "Already committed a lot to something? It's probably a good bet to keep at it," even though this isn't *necessarily* the case, and—as with some regretful gamblers, and government investors in the Concorde— it can lead people astray.

Let's turn next to the dollar auction game. Yale University economist Martin Shubik suggested that you might want to try it some time, maybe during a party at your house . . . so long as you don't mind turning a friendly gathering into a bunch of angry, resentful competitors. Here is the idea. Announce that you will be the auctioneer, and that you are going to auction off a $20 bill. (In the original account, it was $1—hence, the "dollar auction"—but inflation has made $20 more appropriate.) Hold up the $20 and invite your guests to bid for it, the only catch being that both the highest and the next highest bidder have to pay what they have bid to the auctioneer.

Let's imagine what might happen. Someone starts with a bid of $1. Individual 2 then figures, "Heck, I can't let her get away with that, making $19 while I just stand by, doing nothing." So, individual 2 bids, say, $1.50. Not a bad move, since if successful, he stands to make $18.50. But what about individual 1? She is now in a position to lose her original $1 (remember, the highest and next-highest bidder must surrender the amount they bid) and get nothing in return, while individual 2 makes his $18.50 profit. So individual 1 boldly reenters the bidding, offering perhaps $2. At this point, individual 2 may well be moved to make a higher bid yet, or a newcomer, individual 3, might decide to take a shot at it, maybe with a seemingly preemptive bid of $8, since, after all, this could return a profit of $12. But individual 2 won't take kindly to being preempted, since it means a loss for him. And so it goes.

Even though the participants are free to bid or to "pass" at any time, the whole game is a trap, playing upon people's susceptibility to the Concorde fallacy. There are three important moments of truth in this Dollar Auction Game:

1. The decision to enter the bidding at all.
2. Once the total of the bidding surpasses $20 (roughly, after individual bids go past $10) at which point the auctioneer is guaranteed to make a profit, or, another way of looking at it, the highest and next-highest bidders, considered together, are certain to lose. And
3. The really frightening point, after the actual bidding exceeds $20, at which point the bidders are really stuck, each trying to minimize his or her losses rather than seeking to come out ahead.

(Incidentally, if you can't imagine the difficulty and tension generated by playing such a game, then it is especially important that you
not start!)

There are folk warnings about such errors, advising us to refrain
from throwing good money after bad, and so forth. But still it happens.
It has been called the Macbeth dilemma, after Shakespeare's murderous villain who acknowledges the problem, but only after it is too late,
at which point he cannot extricate himself:

I am in blood
Stepped so far that, should I wade no more
Returning were as tedious as go o'er.

Shubik reports that the auction game is "usually highly profitable to
its promoter," since subjects typically bid well over the value of the
reward, once faced with the prospect of losing their earlier bids. It also
tends to evoke strong and negative emotions, making people agitated
and upset, blaming others for *their* irrationality!

There are many day-to-day examples of the dollar auction. Imagine
that you are phoning a business or government agency long distance.
You've been on hold, wasting time and costing money, but as the pre-
recorded message announces, "calls are taken in the order they are
received." You have no idea how much longer you'll have to wait, or
whether the system is malfunctioning and they'll never get to you.
What to do? Should you hold on for another minute, or five, or ten? If
you hang up, you have forfeited whatever benefit you've derived from
waiting as long as you had. And worse yet, you have nothing to show
for the loss. The answer, once again, is that the best strategy is to
remain on the line, or hang up, depending on what is the best use of
your time at any given moment, *not* as a function of how much time
you have already spent on hold. (It can also be helpful when an automatic message informs each caller approximately how much more time
will be required.)

A particularly pernicious example of the dollar auction fallacy
worked its nasty magic during the Vietnam War. "We can't pull out
now," said people who ought to have known better. "We have already

lost ten thousand men, and we cannot allow them to have died in vain." In time, this number became twenty thousand, making it even more difficult to back out. Then thirty thousand, and so on, up to more than fifty thousand.

A big problem is that often we don't realize that we're in a dollar auction until after we've made a bid or two, or three. And by then, given the widely shared susceptibility to the Concorde fallacy, it may be too late.

"Of all the faculties of the human mind," wrote Charles Darwin, "it will, I presume, be admitted that Reason stands at the summit."[17] At the same time, it must also be admitted that like so many other summits, this one is perched perilously close to a long and slippery slope.

"Logical" Vicious Circles

A vicious circle is illogical, especially if it emerges as a solution to what seems to be a rational problem. But in fact, the cool logic of game theory often generates precisely this kind of illogical outcome.

Take, for example, a situation developed in *The Final Problem,* one of Arthur Conan Doyle's Sherlock Holmes stories. It happens that Holmes and his archenemy, Moriarty, are both in London. Holmes is trying to escape from Moriarty, who will kill him if given the opportunity. Moriarty knows that Holmes has boarded a train heading for Dover, but he doesn't know if Holmes is planning to stay on until Dover, or get off at the intermediate stop, Canterbury. If Holmes makes it to Dover, without Moriarty being there, Holmes will be able to escape to continental Europe (this generates Holmes's highest payoff and Moriarty's lowest). But if Moriarty is there, Holmes will be killed (Holmes's lowest payoff and Moriarty's highest). On the other hand, Holmes might choose to get off at Canterbury, in which case, if Moriarty is there, it's another jackpot for Moriarty and loser for Holmes. Or, if Holmes disembarks at Canterbury and Moriarty guesses incorrectly and goes on to Dover, both players receive an intermediate payoff, since Holmes might—or might not—succeed in avoiding Moriarty and getting to Dover eventually.

The situation is a 2 × 2 game, as follows:

		Moriarty gets off at	
		Canterbury	Dover
Holmes gets off at	Canterbury	H loses, M wins	H wins a bit, M loses a bit
	Dover	H wins, M loses	H loses, M wins

In Arthur Conan Doyle's story, Holmes figures that Moriarty will decide to go straight to Dover, trying to keep Holmes from escaping his grasp (that is, to keep Holmes from winning). And so, Holmes gets off at Canterbury and Moriarty goes on to Dover; Holmes eludes Moriarty, in part because Moriarty was trying to avoid his worst outcome (minimax), while Holmes correctly figured that Moriarty would be doing just this.

The Holmes-Moriarty battle of wits is much beloved by game theoreticians, not only because it lends itself to a simple matrix, but also because it was discussed in the founding document of game theory, *Theory of Games and Economic Behavior,* by mathematician John von Neumann and economist Oskar Morgenstern. There is no single best solution to this battle. In a book published in 1928 (coincidentally, the same year von Neumann proved the minimax theorem, which is the primary mathematical underpinning of game theory), Morgenstern pointed out that there could have been an "endless chain" of reasoning between the two old enemies. Thus, after describing Holmes's reasoning—that Moriarty would probably go to Dover, trying to deprive Holmes of his highest payoff, and accordingly, he (Holmes) should get off at Canterbury—Morgenstern asked, "But what if Moriarty had been still more clever, had estimated Holmes's mental abilities better and had foreseen his actions accordingly?" In that case, Moriarty—anticipating that Holmes would actually get off at Canterbury—should also get off at Canterbury.

And what, we may ask, if Holmes had been cleverer yet, and anticipated Moriarty's anticipation of Holmes's anticipation . . . and if Moriarty, in turn, had planned yet further ahead . . . and then if Holmes had . . .

Similar endless loops of sterile reasoning characterize the planning—and anxiety—of nuclear strategists. Imagine a situation of high tension between the United States and the Soviet Union during the Cold War, or between India and Pakistan today. Country A worries that country B might attack preemptively, hoping to gain some advantage from striking first. As a result, country A might feel constrained to preempt that preemption, and beat the other to the punch by attacking first. For its part, country B may have had no intention of striking the first blow, but now it worries that country A might expect that it will, so it feels pressure to preempt country A's preemption, by its own prepreemptive attack. And so on, ad infinitum, ad nauseum, and *ad destructum mutualis.*

Finally, if you think that two-person games can be maddening in their vicious circularity, take a quick look at multiperson games, as some intrepid game theorists have done. It is worth noting, for starters, that a multiperson game cannot be strictly competitive (that is, zero sum), since if it is zero sum for two of the players, then a third cannot have interests that are diametrically opposed to both of them. The prospect therefore exists for coalitions of all sorts: blocs, unions, cliques, etc., which are likely to be peculiarly unstable, and with the potential—as with the Holmes-Moriarty game—of cycling indefinitely.

Consider this one, a variation on the Ultimatum Game examined earlier: Anna, Betty, and Cindy have been offered $100. They can divide it however they want, two ways or three, partitioning the money in any manner they choose, so long as any two of them agree on the distribution. If they don't reach this minimal agreement, no one gets anything. What might happen? For starters, let's assume that they reach this "rational" agreement: to accept $33.33 each (and to donate the remaining penny to charity). But then, what is to stop Anna and Betty from agreeing to divide the $100 among themselves: $50 each, and to exclude Cindy? Such a deal would be entirely rational on their part, and worth an additional $16.67 to each. But suppose Cindy gets wind of the plan and approaches Anna with a better deal yet: Anna can have $60 and she, Cindy, will accept a mere $40, thereby freezing out Betty. (Better for Cindy to get $40 than nothing, and it would be a good deal for Anna, too, since she gets $60 rather than $50.) But now, what is to stop Betty from counteroffering to Anna, proposing that she'll give Anna $70 and accept a mere $30 for herself? (After all, that, too, is better than nothing, and Anna will surely accept.)

At this point, Anna is feeling pretty good until she spots Betty and Cindy whispering together. What might they be up to?

In fact, under these circumstances, there is no limit to the possible machinations. Maybe they will cycle back to an equal, three-way split. But this, once again, will be vulnerable to a two-way division, freezing out the third but benefiting the collaborating pair. Such maneuvering is not limited to money. A literary example comes from George Orwell's vision of the three competing mega-states in his novel, *1984,* in which Eurasia, Oceania, and Eastasia are constantly shifting their alliances, with two ganging up on the third.

In such cases, it is difficult not to conclude that rationality itself has ganged up on its practitioners.

Pas Trop de Zèle!

Now that we've come to the end of this book, I'll make a confession: For a long time I have really loved game theory, and, for about as long, I've hated it. In particular, I've been infuriated by the way nuclear strategists have employed game theory to justify "rational" calculations that come perilously close to bringing about the end of the world. To my mind, their cool-headed logical deductions—with life on earth literally in the balance—have been nothing less than despicable. At the same time, as a biologist and psychologist I have been fascinated to see how the behavior of living things can be modeled and predicted by game theory, especially the insights of "evolutionarily stable strategies." And as a thinking human being, I've been delighted, occasionally baffled, and sometimes exhilarated by the mind games involved.

Game theory is often insightful, frustrating, fun, scary, annoying, challenging, and even—on occasion—irrelevant. The same can probably be said for rationality.

One of Goya's most famous paintings is titled *The Sleep of Reason Produces Monsters.* Monsters, however, arise from many sources, and not just when reason is slumbering and our irrational, unconscious selves have free play. Sometimes, in fact, it is reason itself that generates monstrous outcomes, whether they be hard-eyed nuclear machinations, vicious circles of the Holmes/Moriarty type, or—more commonly— cases of Prisoner's Dilemma, in which we need to be protected from our own excessive rationality. (Another way of putting it: In such cases, irrational players can do better than their more rational counterparts.)

The great French foreign minister Talleyrand, one of the giants of international maneuvering and power politics, had this advice for his diplomats: *"Pas trop de zèle"* (*"not too much zeal"*). Although intended as a warning against emotion and in favor of hardheaded rationality in the conduct of foreign affairs, Talleyrand's injunction applies to devotees of game theory as well. Enjoy the paradoxes and understand the arguments, but beware of too much zeal for game theory itself!

As Buddhist master Thich Nhat Hanh emphasizes, you can point to the moon, but be careful to distinguish your finger from the moon. I offer the following advice in the same spirit: Understand game theory, but don't confuse it with life itself. Have fun with payoff matrices, but please don't lead your life according to precepts derived from them. Be rational, but not too rational. If nothing else, we have already seen that at least on occasion, people need to be protected from their own logical inclinations, no less than from their penchant for irrationality. This is especially true when the rational, game theoretically appropriate route leads to mutual disaster. In short, *Pas trop de zèle.*

Not for nothing have people elevated the teaching of genuinely high-minded moralists—Jesus, St. Francis, Lao Tzu, Confucius, Socrates— who recommend a kind of collective rationality over its more selfish, individualistic alternative. Similarly, it is worth repeating that the payoffs that game theoreticians identify with 2×2 matrices such as Prisoner's Dilemma have a distinctly theological flavor, notably Temptation, Punishment, and Reward. In this company, the fourth payoff—and not coincidentally, the lowest—stands out in its crude undesirability: Sucker. Perhaps this discrepancy can be erased if we see it somewhat differently, although still retaining its first initial: Saint. Lenin—arch-cynic and power-manipulator, who had no use for anything unlikely to yield him a positive payoff—once referred to Tolstoy's ethics as "the sermon of the *yurodivy*" (saintly idiot).

The world needs more *yurodivy,* people who, knowing the payoffs, choose a different kind of rationality, whose zeal is directed toward the betterment of all rather than just themselves.

At the same time, there is also room—indeed, a crying need—for people to think better, more creatively, less restrictively, "out of the box," and, most important, out of love for the world itself as well as, paradoxically, a passion for benevolent rationality. My favorite game theoretician, Anatol Rapoport, who spent decades attempting to connect game theory with the peaceful resolution of human conflicts, has suggested that perhaps game theory's greatest contribution to humanity will be comparatively modest: self-improvement and self-knowledge.[18] He points to the parable in which two sons are told to dig for a treasure in the family vineyard: they did not find any treasure, but their labors greatly enriched the soil.

NOTES

1. THE GAMES WE ALL PLAY

1. Andrew M. Colman, *Game Theory and Its Applications* (Oxford, England: Butterworth Heinemann, 1995).

2. John Maynard Keynes, *General Theory of Employment, Interest, and Money* (London: Macmillan, 1936).

3. Roger Fisher and William Ury, *Getting to Yes* (Boston: Houghton Mifflin, 1982).

4. Stephen Jay Gould, "Losing the Edge" in *The Flamingo's Smile* (New York: Norton, 1985).

5. David P. Barash, *Revolutionary Biology: The New, Gene-Centered View of Life* (New Brunswick, N.J.: Transaction Publishers, 2002).

6. Bertrand Russell, *Human Society in Ethics and Politics* (London: Allen and Unwin, 1954).

7. Martin Shubik, ed., *Readings in Game Theory and Political Behavior* (Garden City, N.Y.: Doubleday, 1954).

8. Steven Brams and Marc Kilgour, "The Truel," *Mathematics Magazine,* 70 (1997): 315–26.

2. MASTERING THE MATRIX

1. James Schellenberg, *Primitive Games* (Boulder, Colo.: Westview, 1990).

2. Thomas Flanagan, *Game Theory and Canadian Politics* (Toronto, Ontario: University of Toronto Press: 1998).

3. R. Duncan Luce and Howard Raiffa, *Games and Decisions* (New York: Wiley, 1957).

4. Anatol Rapoport, "Exploiter, Leader, Hero and Martyr: The Four Archetypes of the 2 × 2 Game," *Behavioral Science*, 12 (1967): 81–84.

5. Anatol Rapoport, *Fights, Games and Debates* (Ann Arbor, Mich.: University of Michigan Press, 1960).

6. Quoted in Ole R. Holsti, Richard A. Brody, and Robert C. North, "Measuring Affect and Action in International Reaction Models: Empirical Materials from the 1962 Cuban Crisis," *Journal of Peace Research*, 1 (1964): 170–89.

7. Quoted in Harold Willensky. *Operational Intelligence* (New York: Basic Books, 1967).

8. O. G. Haywood, Jr., "Military Decision and Game Theory," *Journal of the Operations Research Society of America*, 2 (1954): 365–85.

9. Italo Calvino, *If on a Winter's Night a Traveler* (New York, Harcourt, Brace, Jovanovich, 1981).

3. PRISONER'S DILEMMA AND THE PROBLEM OF COOPERATION

1. William Poundstone, *Prisoner's Dilemma* (New York: Doubleday, 1992).

2. Douglas Muzzio, *Watergate Games: Strategies, Choices, Outcomes* (New York: New York University Press, 1982).

3. Ludwig Rubiner, et al., "Auf Helgoland," trans. David Kramer, in Alexander Mehlmann's *The Game's Afoot: Game Theory in Myth and Paradox* (Providence, R.I.: American Mathematical Society, 2000).

4. Paul E. Turner and Lin Chao, "'Prisoner's Dilemma' in an RNA Virus," *Nature*, 398 (1999): 441–43.

5. James Andreoni and John H. Miller, "Rational Cooperation in the Finitely Repeated Prisoner's Dilemma: Experimental Evidence," *Economic Journal*, 103 (1993): 570–85.

6. Anatol Rapoport, ed., *Game Theory as a Theory of Conflict Resolution* (Boston, Mass.: D. Reidel Publishing Co., 1974).

7. Steven Brams, "Theory of Moves," *American Scientist*, 81 (1993): 562–70.

8. Robert H. Frank, *Passions Within Reason: The Strategic Role of the Emotions* (New York: Norton, 1999).

9. Euripides, *Iphigenia in Tauris* (Chicago, Ill.: Ivan R. Dee, 1997).

10. Juan Carlos Martinez-Coll and Jack Hirshleifer, "The Limits of Reciprocity," *Rationality and Society*, 3 (1991): 35–64; Ken Binmore, *Game Theory and the Social Contract. Vol. I: Playing Fair* (Cambridge, Mass.: MIT Press, 1994).

11. Reinhard Selten, "Re-examination of the Perfectness Concept for Equilibrium Points in Extensive Games," *International Journal of Game Theory*, 4 (1975): 25–55.

12. Jianzhong Wu and Robert Axelrod, "How to Cope with Noise in the Iterated Prisoner's Dilemma," *Journal of Conflict Resolution*, 39 (1995): 183–89.

13. Vivian C. L. Hutson and Glenn T. Vickers, "The Spatial Struggle of Tit-for-Tat and Defect," *Phil. Transcript of the Royal Society of London B*, 348 (1995): 393–404; Regis Ferriere and Richard E. Michod, "Invading Wave of Cooperation in a Spatially Iterated Prisoner's Dilemma," *Proceedings of the Royal Society London B*, 259 (1995): 77–83.

14. Robert Boyd, "The Evolution of Reciprocity When Conditions Vary," in A. H. Harcourt and Frans B. M. de Waal, eds., *Coalitions and Alliances in Humans and Other Animals* (Oxford, U.K: Oxford University Press, 1992); Phillip Kitcher, "The Evolution of Human Altruism," *Journal of Philosophy*, 90 (1993): 497–516.

15. D. V. Kelly, *39 Months* (London: Ernst Benn, 1930).

16. D. Dugdale, *Langemarck and Cambrai* (Shrewsbury, England: Wilding and Son, 1932).

17. Tony Ashworth, *Trench Warfare, 1914–1918: The Live and Let Live System* (New York: Holmes and Meier, 1980).

18. Stair Gillon, *The Story of the 29th Division* (London: Nelson and Sons, n.d.).

19. G. Belton Cobb, *Stand to Arms* (London: Wells Gardner, Darton & Co., 1916).

20. Ian Hay, *The First Hundred Thousand* (London: Wm. Blackwood, 1916).

21. O. Rutter, ed., *The History of the Seventh (Services) Battalion, The Royal Sussex Regiment 1914–1919* (London: Times Publishing Co., 1934).

22. David Grossman, "Trapped in a Body at War with Itself," *New York Times*, August 25, 2001, A23.

23. Manfred Milinski, "TIT-FOR-TAT in Sticklebacks and the Evolution of Cooperation," *Nature*, 325 (1987): 433–35.

24. Eric Fischer, "The Relationship Between Mating System and Simultaneous Hermaphroditism in the Coral Reef Fish, *Hypoplectus nigricans*," *Animal Behaviour*, 28 (1980): 620–33.

25. Rex Warner, trans., *War Commentaries of Caesar* (New York: New American Library, 1960).

26. Martin Mayer, *The Bankers* (New York: Ballantine Books, 1974).

27. Rapoport, *Game Theory*.

28. Anatol Rapoport and A. M. Chammah, *Prisoner's Dilemma: A Study in Conflict and Cooperation* (Ann Arbor, Mich.: University of Michigan Press, 1965).

29. J. E. Alcock and D. Mansell, "Predisposition and Behavior in a Collective Dilemma," *Journal of Conflict Resolution,* 21 (1977): 443–57.

30. H. H. Kelly and A. J. Stahelski, "Errors in Perception of Intentions in a Mixed-motive Game," *Journal of Experimental Social Psychology,* 6 (1970): 379–400.

31. Daniel R. Lutzker, "Internationalism as a Predictor of Cooperative Behavior," *Journal of Conflict Resolution,* 4 (1960): 426–30.

32. J. S. Minas, A. Scodel, D. Marlowe, and J. Rawson, "Some Descriptive Aspects of Two-person Non-zero-sum Games. II," *Journal of Conflict Resolution,* 4 (1960): 193–97.

4. SOCIAL DILEMMAS

1. Robert L. Trivers, "The Evolution of Reciprocal Altruism," *Quarterly Review of Biology,* 46 (1971): 35–57.

2. David P. Barash, "Sociobiology of Rape in Mallards (*Anas platyrhynchos*): Responses of the Mated Male," *Science,* 197 (1977): 788–89.

3. Timothy H. Clutton-Brock, "Reproductive Skew, Concessions and Limited Control," *Trends in Ecology & Evolution,* 13 (1998): 288–92.

4. Plato, *The Republic,* trans. B. Jowett (London: Oxford University Press, 1901).

5. Thomas Hobbes, *The Leviathan* (New York: The Modern Library, 1939).

6. J.-J. Rousseau, *The Social Contract,* trans. G. D. H. Cole (New York: Dutton, 1935).

7. Immanuel Kant, *Lectures on Ethics,* trans. Peter Heath (New York: Cambridge University Press, 1997).

8. A. Mintz, "Non-adaptive Group Behavior," *Journal of Abnormal and Normal Social* Psychology, 46 (1951): 150–59.

9. Adam Smith, *The Wealth of Nations* (New York: W. W. Norton, 1986).

10. Frans B. M. de Waal, *Chimpanzee Politics: Power and Sex Among Apes* (London: Jonathan Cape, 1982).

11. David M. Messick and Christel G. Rutte, in Wim Liebrand, David M. Messick and Henk A. M. Wilke, eds. *Social Dilemmas: Theoretical Issues and Research Findings* (Oxford, England: Pergamon Press, 1992).

12. Garrett Hardin, "The Tragedy of the Commons," *Science,* 162 (1961): 1243–48.

13. Andrew Colman, personal communication.

14. For a good review of how shared resources are actually managed, in a variety of countries and situations, see: Elinor Ostrom, *Governing the Commons* (New York: Cambridge University Press, 1990).

15. Douglas R. Hofstadter, *Metamagical Themas: Questing for the Essence of Mind and Pattern* (New York: Basic Books, 1985).

16. L. Conrad and T. J. Roper, "Group Decision-making in Animals," *Nature,* 421 (2003): 155–58.

17. For an engaging analysis of a situation in which game theory fails to consider the larger social context, see: Martin A. Nowak, Karen M. Page, and Karl Sigmund, "Fairness Versus Reason in the Ultimatum Game," *Science,* 289 (2003): 1773–75.

18. Mancur Olson, *The Logic of Collective Action* (Cambridge, Mass.: Harvard University Press, 1965).

19. C. G. Rutte, H. Wilke, and D. Messick, "Scarcity or Abundance Caused by People or the Environment as Determinants of Behavior in the Resource Dilemma," *Journal of Experimental Social Psychology,* 23 (1987): 208–16.

20. Kipling D. Williams, J. M. Jackson, and Steven J. Karau, "Collective Hedonism: A Social Loafing Analysis of Social Dilemmas," in *Social Dilemmas,* ed. David A. Schroeder (Westport, Conn.: Praeger, 1995).

21. Brian Bertram, "Vigilance and Group Size in Ostriches," *Animal Behaviour,* 28 (1980): 278–86.

22. Ernst Fehr and Simon Gachter, "Altruistic Punishment in Humans," *Nature,* 415 (2002): 137–40.

23. James K. Rilling, D. A. Gutman, et al., "A Neural Basis for Social Cooperation," *Neuron,* 35 (2002): 395–405.

24. R. J. Hollingdale, ed., *A Nietzsche Reader* (New York: Penguin, 1977).

25. Roderick M. Kramer and Lisa Goldman, "Helping the Group or Helping Yourself?," in David A. Schroeder, ed., *Social Dilemmas.*

26. Ibid.

27. Alphons J. C. van de Kragt, John M. Orbell, and Robyn M. Dawes, "The Minimal Contributing Set as a Solution to Public Goods Problems," *American Political Science Review,* 77 (1983): 112–22.

28. Sanford L. Braver, "Social Contracts and the Provision of Public Goods," in David A. Schroeder, ed., *Social Dilemmas.*

29. Nathaniel Philbrick, *In the Heart of the Sea: The Tragedy of the Whaleship Essex* (New York: Viking, 2000).

5. GAMES OF CHICKEN

1. Quoted in Ole R. Holsti, Richard A. Brody, and Robert C. North, "Measuring Affect and Action in International Reaction Models: Empirical Materials from the 1962 Cuban Crisis," *Journal of Peace Research,* 1 (1964): 170–89.

2. Cited by Thomas Schelling, *Arms and Influence* (New Haven, Conn.: Yale University Press, 1966).

3. "Thinking Ruins Character" quoted in Alexander Mehlmann, *The Game's Afoot: Game Theory in Myth and Paradox* (Providence, R. I.: American Mathematical Society, 2000).

4. David P. Barash and Judith Eve Lipton, *The Myth of Monogamy: Fidelity and Infidelity in Animals and People* (New York: Henry Holt/Times Books, 2002).

5. Ibid.

6. Anatol Rapoport, *Fights, Games and Debates* (Ann Arbor, Mich.: University of Michigan Press, 1960).

7. Morton Deutsch and R. M. Krauss, "The Effect of Threat upon Interpersonal Bargaining," *Journal of Abnormal and Normal Social Psychology,* 61 (1960): 181–89.

8. H. R. Haldeman and Joseph DiMona, *The Ends of Power* (New York: Times Books, 1978).

9. Herman Kahn, *On Escalation* (New York: Praeger, 1965).

10. Colin Camerer and Richard H. Thaler, "Ultimatums, Dictators, and Matters," *Journal of Economic Perspectives,* 9 (1995): 209–19; Elizabeth Hoffman, Kevin McCabe, and Vernon Smith, "Behavioral Foundations of Reciprocity: Experimental Economics and Evolutionary Psychology," *Economic Inquiry,* 36 (1998): 335–52.

11. Andrew Colman, personal communication.

12. Lewis Fry Richardson, *Arms and Insecurity A Mathematical Study of the Causes and Origin of War* (Pittsburgh, Pa., Boxwood Press, 1960).

13. John C. Harsanyi, "Game Theory and the Analysis of International Conflict," *Australian Journal of Politics and History,* 11 (1965): 292–304.

14. Geoffrey Miller, *The Mating Mind* (New York: Doubleday, 2000).

6. ANIMAL ANTICS

1. Michio Hori, "Frequency-dependent Natural Selection in the Handedness of Scale-eating Cichlid Fish," *Science,* 260 (1993): 216–19.

2. Barry Sinervo and Curtis M. Lively, "The Rock-Paper-Scissors Game and the Evolution of Alternative Male Strategies," *Nature,* 380 (1996): 240–43.

3. Geoffrey A. Parker, "Evolutionarily Stable Strategies," in John R. Krebs and Nicholas B. Davies, eds. *Behavioural Ecology: An Evolutionary Approach* (Oxford, England: Basil Blackwell, 1984).

4. R. A. Fisher, *The Genetical Theory of Natural Selection* (New York: Dover, 1930 [1958]).

5. John Maynard Smith and George R. Price, "The Logic of Animal Conflict," *Nature,* 246 (1973): 15–18.

6. Robert M. Fagen, "When Doves Conspire: Evolution of Nondamaging Fighting Tactics in a Nonrandom Encounter Animal Conflict Model," *The American Naturalist,* 115 (1980): 858–59.

7. Ken Binmore, *Playing Fair* (Cambridge, Mass.: MIT Press, 1994).

8. Ken Binmore, *Fun and Games* (Lexington, Mass.: D.C. Heath and Co., 1992).

9. Alan Grafen, "The Role of Divisively Asymmetric Contests: Respect for Ownership and the Desperado Effect," *Animal Behaviour,* 35 (1987): 462–67.

10. Ken Yasukawa, "A Fair Advantage in Animal Confrontations," *New Scientist,* 1 (1979): 366–68.

11. N. Davies, "Territorial Defense in the Speckled Wood Butterfly: The Resident Always Wins," *Animal Behaviour,* 26 (1978): 138–47.

12. P.-O. Wickman and Christer Wiklund, "Territorial Defense and Its Seasonal Decline in the Speckled Wood Butterfly (*Parage aegeria*)," *Animal Behaviour,* 31 (1983): 1206–16.

13. Hans Kummer, *Primate Societies* (Chicago, Ill.: Aldine Atherton, 1971).

14. J. Wesley Burgess, "Social Spiders," *Scientific American,* 234 (1976): 100–6.

15. Richard Dawkins, "Good Strategy or Evolutionarily Stable Strategy?," in George Barlow and James Silverberg, eds., *Sociobiology: Beyond Nature/Nurture?* (Boulder, Colo.: Westview Press, 1980).

16. See David P. Barash and Judith Eve Lipton, *Gender Gap: The Biology of Male-Female Differences* (New Brunswick, N.J.: Transaction Publishers, 2002).

17. John Maynard Smith, *Evolution and the Theory of Games* (New York: Cambridge University Press, 1982).

18. Peter Schuster and Karl Sigmund, "Coyness, Philandering, and Stable Strategies," *Animal Behaviour,* 29 (1981): 186–92.

19. Randy Thornhill, "*Panorpa* (Mecopters: Panorpidae) Scorpionflies: Systems for Understanding Resource-defense Polygyny and Alternative Male Reproductive Efforts," *Annual Review of Ecology and Systematics,* 12 (1981): 355–86.

20. David P. Barash, "Mate Guarding and Gallivanting by Male Hoary Marmots (*Marmota caligata*)," *Behavioral Ecology and Sociobiology,* 9 (1981): 187–93.

21. Stephen M. Shuster, "The Reproductive Behaviour of A-, B- and G-male Morphs in *Paracerceis sculpta:* A Marine Isopod Crustacean," *Behaviour,* 121 (1992): 231–58.

22. Stephen M. Shuster and Michael J. Wade, "Equal Mating Success Among Male Reproductive Strategies in a Marine Isopod," *Nature,* 350 (1991): 608–10.

23. Mart R. Gross, "Sneakers, Satellites and Parentals: Polymorphic Mating Strategies in North American Sunfishes," *Zeitschrift fur Tierpsychologie,* 60 (1982): 1–26.

7. THOUGHTS FROM THE UNDERGROUND

1. Ken Binmore, *Fun and Games* (Lexington, Mass.: D.C. Heath and Co., 1992).

2. Rebecca West, *Black Lamb and Grey Falcon* (New York: Viking, 1940).

3. Anthony Downs, *An Economic Theory of Democracy* (New York: Harper & Row, 1957).

4. David Hume, *A Treatise of Human Nature* (New York: Oxford University Press, [1739] 1978).

5. Will Durant, *Rousseau and Revolution* (New York: Macmillan, 1967).

6. Stephen Crane, *The Black Riders and Other Lines* (New York: Alfred A. Knopf, 1895).

7. L. G. Humphreys, "Acquisition and Extinction of Verbal Expectations in a Situation Analogous to Conditioning," *Journal of Experimental Psychology,* 25 (1939): 294–301.

8. Particularly good collections can be found in Daniel Kahneman and Amos Tversky, eds., *Choices, Values, and Frames* (New York: Cambridge University Press, 2000).

9. Daniel Kahneman and Amos Tversky, "Choices, Values, and Frames," in Kahneman and Tversky.

10. This problem was first suggested by Daniel Ellsberg, "Risk, Ambiguity and the Savage Axioms," *Quarterly Journal of Economics,* 75 (1961): 643–69; further examples can be found in Colin F. Camerer and Martin W. Weber, "Recent Developments in Modelling Preferences: Uncertainty and Ambiguity," *Journal of Risk and Uncertainty,* 5 (1992): 325–70.

11. Richard H. Thaler, "Mental Accounting Matters," in Kahneman and Tversky.

12. William Shakespeare, *A Midsummer Night's Dream* (New York: Dover Publications, 1992), act II, scene ii.

13. Herbert Simon, *Models of Man* (New York: John Wiley & Sons, 1957).

14. Leda Cosmides, "The Logic of Social Exchange," *Cognition,* 31 (1989): 187–276.

15. Martin Shubik, "The Dollar Auction Game: A Paradox in Noncooperative Behavior and Escalation," *Journal of Conflict Resolution,* 9 (1971): 109–11.

16. Richard Dawkins and R. T. Carlisle, "Parental Investment, Mate Desertion and a Fallacy," *Nature,* 262 (1976): 131–32.

17. Charles Darwin, *The Descent of Man and Selection in Relation to Sex,* 2nd edition (London: Murray, 1871).

18. Anatol Rapoport, *Fights, Games and Debates* (Ann Arbor, Mich.: University of Michigan Press, 1960).

ACKNOWLEDGMENTS

I especially want to thank my former editor and now agent, John Michel, who encouraged me throughout this project, and whose advice I continue to cherish. Robin Dennis, at Times/Holt, proved herself to be both eagle-eyed and intellectually sprightly, also providing the in-house support that every author hopes for and too few receive. The following "game theory dudes"—as my teenaged daughter calls them—read the manuscript and made helpful comments: Robert Axelrod, Steve Brams, Andrew Colman, Lee Dugatkin, and Herb Gintis. I am grateful to each, and apologize now for not always following their advice. Special thanks to the honors students who participated in my game theory seminar at the University of Washington, and who benevolently forced me to clarify much that was obscure, and to discard much (although probably not all!) that was "over the top." And, of course, my unending gratitude to my family for their love, and for putting up with my games . . . even as I do the same with theirs!

INDEX

ABOUT THE AUTHOR

David P. Barash, Ph.D., is a zoologist and currently professor of psychology at the University of Washington in Seattle. One of the early contributors to the field of sociobiology, he continues to conduct field studies of animal behavior, particularly the evolution and ecology of social systems among free-living animals such as marmots and pikas. His current work has been directed to understanding the underlying evolutionary factors influencing human behavior and evolutionary psychology. Since the early 1980s he has also been active in researching, promoting, and practicing the field of peace studies, focusing on how biology affects behavior, including male-female differences, reproductive strategies, and violence among living things.

Barash has written nearly two dozen books, including *The Myth of Monogamy* and *Making Sense of Sex* with Judith Eve Lipton, *Revolutionary Biology*, and *The Mammal in the Mirror*, as well as popular articles in *Playboy*, *Psychology Today*, and *The New York Times*.

He lives in Redmond, Washington.